More Mathematical Byways

Recreations in Mathematics

Series Editor
David Singmaster

More
Mathematical
Byways

Hugh ApSimon

Oxford   New York
OXFORD UNIVERSITY PRESS
1990

Oxford University Press, Walton Street, Oxford OX2 6DP

Oxford New York Toronto
Delhi Bombay Calcutta Madras Karachi
Petaling Jaya Singapore Hong Kong Tokyo
Nairobi Dar es Salaam Cape Town
Melbourne Auckland

and associated companies in
Berlin Ibadan

Oxford is a trade mark of Oxford University Press

British Library Cataloguing in Publication Data
ApSimon, Hugh, 1926–
    More Mathematical byways.—(Recreations in mathematics)
    1. Mathematical puzzles
    I. Title   II. Series
    793.7'4
ISBN 0–19–217777–X

Library of Congress Cataloging in Publication Data
ApSimon, Hugh.
    More Mathematical byways/Hugh ApSimon.
        p.   cm.—(Recreations in mathematics)
        1. Mathematical recreations.   I. Title   II. Title: More Mathematical
    byways.   III. Title: More Mathematical byways.   IV. Series.
    QA95.A66 1990   793.7'4—dc20   89–255549
ISBN 0–19–217777–X

Typeset by Cotswold Typesetting Ltd.
Printed in Great Britain by
Biddles Ltd, Guildford and King's Lynn

To
HELEN
sometime Mathematician
(and fascinating problem)

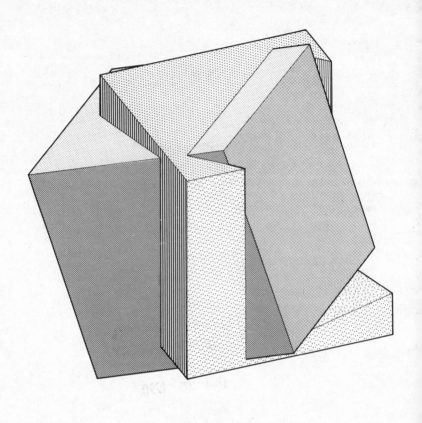

PREFACE

When *Mathematical Byways in Ayling, Beeling, and Ceiling* was published, in 1984, the subsequent advice that I received from reviewers and correspondents was fairly wide-ranging.

One reviewer took me severely to task for my Male Chauvinism: my three characters should—she said—have been Tom, Dick, and Harriet rather than Tom, Dick, and Harry. I take her point: in this book Tom, Dick, and Harry are still with us—but they have been joined by Anne and Belinda. Anne is a better mathematician than any of the others. (And I've a suspicion that she's also better than I am—which is a little disconcerting.)

Three correspondents felt the problems to be too uniform in difficulty: in this book three (at least) of the problems are harder than any in the earlier book; at the other end of the scale, 'Alphametics' may provide light relief. (And there's one problem that is no problem at all.)

Two correspondents complained that (once they knew how) all the problems except one ('Counterfeit Coins III') could be solved with little slog. Muttering to myself that there's no accounting for tastes, I have included in this book a few problems where considerable slog seems to be necessary to achieve the solution. I am a little uneasy about doing so—but I comfort myself with the thought that I am providing additional challenges to my readers: to find significantly shorter solutions than I have been able to do!

And lastly, in answer to several correspondents, I *don't* pick on a mathematical tool and then try to find a problem with which to illustrate it. For me, it's the problem that comes first; and in the early stages of attempted solution I've little idea about how difficult it's going to be—or what mathematical tools are going to be involved. (One consequence is that most 'problems' finish up in the waste-paper basket, as being far too easy or far too difficult.) The problems in this book have—so far—avoided the waste-paper basket. I hope that you enjoy them.

Frimley, Surrey   H. ApS.
1989

CONTENTS

INTRODUCTION

The general arrangement of each chapter (though no chapter contains *all* of the following sections) is:

(i) *'Problem'*: a specific problem (or problems) set, in most cases, in the context of one or more of the villages of Ayling, Beeling, and Ceiling, and involving some or all of Tom, Dick, Harry, Anne, and Belinda. (There are a few exceptions—the chapter on 'Alphametics', for example, does not involve any of the villages or any of their inhabitants. And there's one chapter—'Potential Pay'—that doesn't really pose a problem at all.)

(ii) *'General solution'*: an investigation of the more general problem of which the original problem is a particular example. (But this section is often fused with the next one.)

(iii) *'Particular solution'*: the application of the general solution to the original specific problem, or problems.

(iv) *'Composer's problem'*: a general chat about the difficulties I had in setting the problem, or in laying out the solution—or about how I came across the problem in the first place. There's a *Composer's problem* section in every chapter except 'Benedict's Birthday'. (It's not that I had no problems with setting 'Benedict': I had several. But they were very *boring* problems.)

(v) *'Extension'*: a new problem—usually a more difficult one—that has been suggested to me by the original problem; a problem that I have not been able to solve (at the time of writing this introduction). I am leaving the solution of the *Extension* problems to you.

(vi) *'Appendix'*: that part of the solution of the problem that (for completeness) needs to be included—but which is unfortunately rather turgid, and so needs to be relegated to an inconspicuous position.

The five 'Alphametic' problems have been grouped together in one chapter, principally because each needs a preliminary statement of just what an Alphametic *is*, and it would have been a boring extravagance to make that statement five times over. But I do *not* suggest that you tackle Alphametics one after the other, without pause for breath between them: perhaps they are best used as light relief between more serious or more difficult problems.

1

DOUBLE DEALING

Problem

The three villages of Ayling, Beeling, and Ceiling lie in a plain. There used to be two other villages—Wheeling and Dealing—in the same plain, but nothing now remains above ground to show where they were (though everyone is convinced that a treasure was buried in the centre of Dealing shortly before the two villages were obliterated).

Dick walked the seven-and-a-half miles (well, almost ten yards more, if you want to be pedantic) from Ceiling to Beeling—it's a straight path between the two—in a state of pleasurable excitement, to meet Tom at the Beeling Bistro and tell him: 'I've just found this old book: according to it, Wheeling was as far from both Ceiling and Ayling as Ceiling is from Ayling; and Dealing was as far from both Wheeling and Beeling as Wheeling was from Beeling. So let's go and get that treasure.'

'Hang on,' said Tom. 'It would be useful to know a bit more: for example, if one went straight from here to Wheeling, would Dealing be somewhere off to the left or somewhere off to the right?'

'Good point,' said Dick. 'The book's a bit obscure; but I've worked out that it would be somewhere off to the left if, when one went straight from Ayling to Wheeling, Ceiling was somewhere off to the right—and somewhere off to the right if, when one went straight from Ayling to Wheeling, Ceiling was somewhere off to the left.'

'That helps,' said Tom.

'And', added Dick, 'there was a bit in the book that said how far Dealing was from Ayling—but somebody's cut that bit out, so I don't know.'

'I think we could work it out,' said Tom, 'but all the same . . .'

'I'm off,' said Dick, and left in his search for Dealing and the treasure. Tom left a little later, on the same quest. Each followed the book faithfully—but they finished up, digging furiously, at two points well apart from each other.

How far apart?
And how far was each of them from Ayling?

General solution

We do *not* know where Ayling, Beeling, and Ceiling are in relation to each other (even though Tom and Dick presumably do), and at first sight this is a major obstacle. But at least we can assume that they are *somewhere*, as in Diagram 1.1 (in which A ≡ Ayling, B ≡ Beeling, C ≡ Ceiling).

In what follows I present the argument in only the 'Diagram 1.1' case: if the shape of the triangle ABC were different, the *argument* would be the same— but the *presentation* might need to be altered (involving the replacement of '+' by '−', 'α' by 'π − α' and so on) at a number of points. But to give all the possible cases would be distractingly lengthy. (The rightly suspicious reader is invited to start with a significantly different shape for ABC, and to work it through for himself. Or to use trigonometry.)

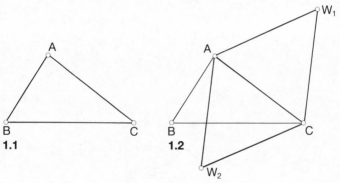

Diagrams 1.1–1.2

We are told that W (≡ Wheeling), C, and A form an equilateral triangle; so there are just two possible positions for W: label them W_1 and W_2 (see Diagram 1.2).

We are then told that D (≡ Dealing), B, and W form an equilateral triangle. Given B and W_1, there are at first sight two possible positions for D—but only one of them meets Dick's point (in answer to Tom's question) that, if C is somewhere off to the right when one goes straight from A to W, then D is somewhere off to the left when one goes straight from B to W. So we take that one, and label it D_1.

Similarly, given B and W_2, there is only one possible position for D: we label it D_2.

D_1, D_2, W_1, and W_2 are shown in Diagram 1.3, which we now look at more closely (and develop in Diagram 1.4).

First:

$$D_1 W_1 = B W_1 \quad (BD_1 W_1 \text{ is equilateral}),$$
$$A W_1 = C W_1 \quad (CAW_1 \text{ is equilateral}),$$
$$A\widehat{W_1}D_1 = B\widehat{W_1}C \quad (\text{each is } 60° - A\widehat{W_1}B).$$

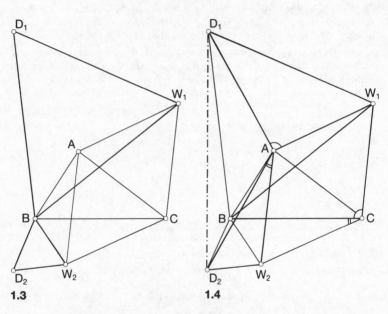

Diagrams 1.3 – 1.4

So the triangles $D_1 W_1 A$ and $B W_1 C$ are congruent. Hence

$$AD_1 = BC, \tag{1a}$$

and

$$D_1\widehat{A}W_1 = B\widehat{C}W_1. \tag{1b}$$

Similarly,

$$AD_2 = BC, \tag{2a}$$

and

$$D_2\widehat{A}W_2 = B\widehat{C}W_2. \tag{2b}$$

From (1b), (2b) we have

$$D_1\widehat{A}W_1 + D_2\widehat{A}W_2 = B\widehat{C}W_1 + B\widehat{C}W_2$$
$$= 120°;$$

so (looking at the cluster of angles surrounding A)

$$D_1\widehat{A}D_2 = 120°. \tag{3}$$

By (1a), (2a), we have

$$AD_1 = AD_2 = BC. \tag{4}$$

So (D_1AD_2 is isosceles) by (3), (4)

$$D_1D_2 = \sqrt{3} \times BC. \tag{5}$$

Particular solution

We are told—in the particular problem—that BC is 'seven and a half miles (well, almost ten yards more)'.

So, by (4), that is the answer to the second question ('How far was each of them from Ayling?').

By (5) we have that the answer to the first question ('How far apart?') is $\sqrt{3}$ times that distance. So: 13 miles.

Composer's problem

I had composed 'To The Point':[1] I was drawing the diagrams for it. D_1D_2 looked perpendicular to BC.[2] Two minutes' coordinate geometry verified that D_1D_2 is indeed perpendicular to BC. It also presented me with the fact, which previously I had not even suspected, that $D_1D_2 = \sqrt{3} \times BC$, quite irrespective of where A is. Once I knew the answer, its proof by pure geometry[3] was short, simple, and 'obvious'. (Obvious, that is, when one knows the answer, or at least that there's an answer worth looking for.)

So I had a viable Problem. But there was a difficulty attached to it: what to do about that $\sqrt{3}$? I prefer problems that have integer data to

---

[1] The next problem (though now, of course, with a completely different narrative).
[2] In the notation of 'Double Dealing'. In the notation of 'To The Point' I in fact said to myself 'ZZ' looks perpendicular to AB.'
[3] I am told that the correct term is 'synthetic geometry' (in antithesis to 'analytic geometry'). I think 'pure geometry' to be more emotively accurate. So I use it.

have integer solutions. One way of bringing that about would have been to give the data in terms of an area ('the area of ABC is 40 square miles'), and then to ask for the area of a triangle ED_1D_2 similar to ABC (it would have been 120 square miles); but that would have distracted attention from what the problem was really about. (And in any case, I couldn't think of a problem that brought a triangle ED_1D_2 naturally into the discussion.)

Another way would have been to give a datum ('BC = 10864 yards') that led to a number of yards for D_1D_2 that was close to an integer[4] (BC = 10864 implies that $D_1D_2 \cong 18817 - 0.000033$), and then to pretend that '0.000033' didn't exist: I am glad to say that I resisted the temptation.

I eventually decided that, since we would (reluctantly) be dealing in approximations anyhow, it was best to stress the fact right from the beginning of the problem statement: 'seven-and-a-half miles (well, almost ten yards more, if you want to be pedantic)'.[5] But I am still not entirely happy: I have two extension questions.

Extension

1.   Compose a problem that illustrates

$$D_1D_2 = \sqrt{3} \times BC,$$

without involving $\sqrt{3}$ (or any other irrational number) in either the data or the answer.

2.   Find a short proof, using only pure geometry, that D_1D_2 is perpendicular to BC.

---

[4] $\sqrt{3} = 1 + \frac{1}{1+} \frac{1}{2+} \frac{1}{1+} \frac{1}{2+} \frac{1}{1+} \frac{1}{2+} \ldots$. Successive convergents are $\frac{1}{1}, \frac{2}{1}, \frac{5}{3}, \frac{7}{4}, \frac{19}{11}, \frac{26}{15}, \frac{71}{41}, \frac{97}{56}, \frac{265}{153}, \frac{362}{209}, \frac{989}{571}, \frac{1351}{780}, \frac{3691}{2131}, \frac{5042}{2911}, \frac{13775}{7953}, \frac{18817}{10864}, \ldots$.

[5] Here I am using the $\frac{26}{15} (= \frac{13}{7\frac{1}{2}})$ approximation to $\sqrt{3}$. If D_1D_2 is 13 miles, then BC is $\frac{13}{\sqrt{3}}$ miles, which—to a much closer approximation—is $\cong 7\frac{1}{2}$ miles + 9.774 yards.

2

TO THE POINT

Problem

Next day[1] at the Ayling Arms Dick told Harry all about the treasure hunt—and when Tom and Anne came in he was more than ready to start all over again to tell Anne.

'Wait a minute,' said Harry: 'don't spoil things. You've given me an idea for a construction problem, and I'd like to try it on Anne. You two keep quiet. Anne: here's a sheet of paper; I've marked on it Ayling, Beeling, and Ceiling—well, actually, points A, B, and C, for short. Here's a straight-edge, a pair of compasses, and a pencil. I want you to construct four points—let's call them P, X, Y, Z—such that PX = BC, PY = CA, PZ = AB, and XYZ is an equilateral triangle. What's the least number of lines—straight lines or arcs of circles—that you need to draw in order to do it? I'll give you ten minutes. Bet you a drink on it. Fair enough?'

Eight minutes later Anne told Harry the answer; and showed him the paper, with four points labelled P, X, Y, Z (which he'd asked for)—and two extra points labelled Y', Z' (which he hadn't asked for).

'I thought I'd draw a few more lines,' she said, 'and give you *both* possible sizes of triangle XYZ. XY'Z' is the second one. So you can make that drink a double, Harry.'[2]

What is the answer to Harry's question?
How many lines did Anne draw altogether?

---

[1] The day after the search for buried treasure described in 'Double Dealing'.

[2] Rumour has it that she also said: 'And you know what you can do with that straight-edge.' This is indignantly denied by Tom. Nevertheless, it's a clue.

Solution

Anne, faced with Diagram 2.1(i), drew six lines altogether—all of them arcs of circles—to give Diagram 2.2(i).[3] In detail:

Label X: the point B.

(This choice is arbitrary: Anne could equally well choose to label X the point C. But, the choice having been made, from now on there is only one possible construction.)

Label P: the point C.
Draw Arc(1): with centre C and radius CA.

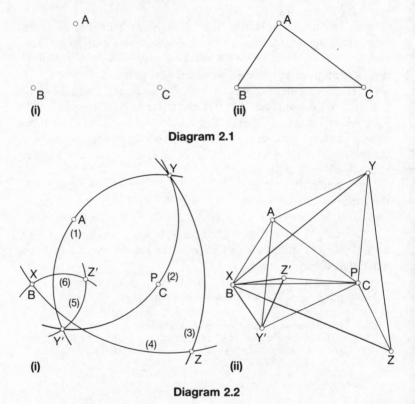

Diagram 2.1

Diagram 2.2

[3] Diagrams 1(ii) and 2(ii) have been drawn by me, not by Anne; their purpose is simply to illustrate the geometry.

Draw Arc(2): with centre A and radius CA.
Label Y, Y': the meeting-points of Arc(1) and Arc(2).
Draw Arc(3): with centre B and radius BY.
Draw Arc(4): with centre Y and radius BY.
Label Z: that meeting-point of Arc(3) and Arc(4) for which A→C→Y and X→Y→Z are in opposite senses.[4]

(Having drawn four lines, Anne by now has the four points that Harry had asked for.)

Draw Arc(5): with centre B and radius BY'.
Draw Arc(6): with centre Y' and radius BY'.
Label Z': that meeting-point of Arc(5) and Arc(6) for which A→C→Y' and X→Y'→Z' are in opposite senses.

(This completes the construction—and Anne has not needed to use the straight-edge at all.[5])

The proof that this construction gives the required answer is, in outline:

By construction: XYZ is equilateral, PX = BC, and PY = CA. PYZ and AYB are congruent (PY = AY, YZ = YB, $P\widehat{Y}Z = A\widehat{Y}B$), so PZ = AB. Similarly with Y', Z' in place of Y, Z.

Composer's problem

In March 1984 I proposed a possible Braintwister to Douglas Barnard:[6]

I provide you with a pair of compasses, a straight-edge, and a piece of paper on which are drawn three straight line segments (lengths a, b, c, say). I ask you to construct four points A, B, C, P such that ABC is an equilateral triangle and AP = a, BP = b, CP = c. Supposing that it is possible at all, what is the least number of operations that you need to do it?

His reply included:

. . . I then smashed my totally useless straight-edge and chewed up the pieces in a rage.

---

[4] That is, of journeys A→C→Y and X→Y→Z, one is clockwise and the other is anti-clockwise.
[5] Which gives some substance to the rumour reported in footnote 2.
[6] Published (slightly edited) as Braintwister 340 in Douglas's column in the *Daily Telegraph*.

(It *may* be that that was what Anne had in mind—if indeed she made her rumoured remark to Harry.)

I have now modified that problem in two ways, and for two reasons:

(i) The construction (supposing that it was possible at all) involved using the pair of compasses not only to draw five arcs, but also to transfer two lengths from one part of the paper to another.[7] There are those who would say that that is not a permissible usage in 'straight-edge and pair of compasses' constructions. So to avoid possible controversy I now give the lengths as the sides of a triangle rather than as separate segments. (Doing so, incidentally, removes the need for the 'supposing that it is possible at all'.)

(ii) The original problem asked for the construction of just *one* equilateral triangle with the required property: it ignored the fact that there are always[8] *two* different equilateral triangles (different in size, that is) having the property. It seemed desirable that the point should be made, fairly forcefully.[9]

'To The Point' may be a little too easy a problem—at least for those who have solved (or read) 'Double Dealing'. So I have inserted a small trap.

If you look at Diagram 1.4 of 'Double Dealing' you may feel that all you have to do is to relabel various points in order to get an appropriate diagram for 'To The Point'. Then clearly P is going to be at A; X, Y, Z must be at D_1, W_1, B (in that order, since Harry requires that PX = BC, PY = CA, PZ = AB); and X, Y', Z' must be at D_2, W_2, B (in that order, for the same reason). But X can not be both at D_1 and at D_2.

(There would have been no difficulty at all if Anne had said: 'X'Y'Z is the other one.' But she didn't.)

Give yourself a bonus mark if you realized the need for a different construction if Anne's 'XY'Z' is the other one' is to be satisfied!

---

[7] The first two arcs transfer two of the segments to form a triangle with the third segment—and one of these arcs is also the first arc in Anne's construction (which carries on from there).

[8] Well, almost always. (The only exception is when the triangle formed by the three segments has zero area.)

[9] Anne: 'So you can make that drink a double, Harry.'

3

FROM THE POINT

Problem

'Harry,' said Tom, 'let's take it the other way round:[1] suppose that XYZ is an equilateral triangle, and that P is a point inside it. If I were to tell you the lengths PX, PY, and PZ, could you calculate—not just draw and measure, of course, but actually calculate—the length of side of XYZ?'

There was a considerable period of silence (broken only by mutters of 'a bit too much like school'); a lot of paper was used; some looked at Anne's construction (and others didn't); until eventually:

'Yes,' said Harry, 'and algebraically it's rather interesting; but numerical answers would be horrid, wouldn't they? Square-root signs all over the place.'

'Not more than a couple,' said Tom, 'and not even those if I chose friendly values for PX, PY, and PZ in the first place.'

'But Tom,' said Dick, 'there'd be *two* possible answers—just as in the construction Anne did.'

'No,' said Tom, 'though I must admit that it's as easy to go looking for both solutions in Anne's construction as it is to go looking for just one of them; but remember that I told you that P is *inside* XYZ, so only one of those two solutions—the larger one—could be an answer to *my* problem.'

'Got one!' yelled Anne: 'Look; if you told us that PX, PY, PZ were 3, 5, 7 (inches, or yards, or whatever—but let's forget about the units), then Dick's two solutions would be 8 and $\sqrt{19}$; 8 is the larger of the two; so 8 would be the answer to your problem.'

'Would it?' said Tom. 'Where would P be?'

'Bother,' said Anne, 'on one of the sides of XYZ. Won't that do?'

'NO,' said Tom. 'P is *inside* XYZ—and inside means inside.'

'Double bother,' said Anne.

---

[1] The other way round from 'To The Point' (which gives Anne's construction).

'Hang on,' said Harry. 'We can twiddle Anne's numbers round a bit. Suppose that you told us that PX, PY, PZ were 3, 7, 8; then you must get 5 as one of Dick's two solutions, and the other would be—wait a minute: $\sqrt{97}$. That's all right, isn't it?'

'Y-e-s,' said Tom, 'but isn't that one of the horrid numbers that you were objecting to in the first place?'

'Well; it's not all *that* horrid,' said Harry; but he said it without conviction.

'All right, Tom,' said Dick. 'You're going to give us whole-number values for PX, PY, PZ; they're going to lead to a whole-number value for the side-length of XYZ, which is an equilateral triangle; and P is definitely inside XYZ. Have I got it right?'

'Absolutely,' said Tom, 'and I'll tell you in addition that that whole-number side-length of XYZ is the smallest for which it can be done. But I've changed my mind about telling you the values of PX, PY, PZ: why don't *you* tell *me* what they are?'

'That', said Anne firmly, 'is too difficult. It could take ages—or a computer.'

'Very well,' said Tom. 'I'll give you a hint. Strictly speaking, you don't need it; but it so happens that in this particular case the values of PX, PY, PZ are in arithmetic progression.'

'Um,' said Anne, 'that helps rather a lot.'

So: what is the smallest whole-number side-length of an equilateral triangle XYZ for which PX, PY, PZ (P inside XYZ) are also whole numbers? And what are those numbers?

Solution

1 *First part—the algebra*

1.1 XYZ is an equilateral triangle and P is a point inside it. We want to calculate the length YZ (call it d) in terms of the lengths PX, PY, PZ (call them a, b, c, respectively).

1.2 We can use an 'Anne-type' construction[2] (in reverse). Take a point, Q, such that PQZ is equilateral and Q is on the opposite side of PZ to X. (See Diagram 3.1.) Then PXZ and QYZ are congruent (PZ = QZ, XZ = YZ, $\widehat{PZX} = \widehat{QZY}$), and so PX = QY. So PQY is a triangle with sides a, b, c (a opposite P, b opposite Q, c opposite Y).

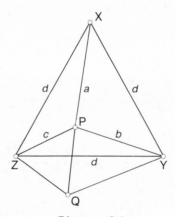

Diagram 3.1

Now P is inside XYZ, so $\widehat{QPY} < 2\pi/3$. But

$$QY^2 = PY^2 + PQ^2 - 2PY.PQ \cos \widehat{QPY}.$$

So

$$a^2 < b^2 + bc + c^2. \tag{1a}$$

Similarly,

$$b^2 < c^2 + ca + a^2, \tag{1b}$$

and

$$c^2 < a^2 + ab + b^2. \tag{1c}$$

---

[2] See 'To The Point'.

1.3   Writing $\widehat{QPY} = \theta$, we develop Diagram 3.2 from Diagram 3.1 and have

$$a^2 = b^2 + c^2 - 2bc \cos \theta \tag{2}$$

and

$$d^2 = b^2 + c^2 - 2bc \cos(\theta + \pi/3). \tag{3}$$

Remembering that

$$\cos(\theta + \pi/3) = \tfrac{1}{2} \cos \theta - \tfrac{1}{2}\sqrt{3} \sin \theta,$$

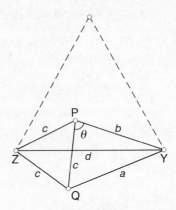

Diagram 3.2

we subtract (2) from twice (3), to give

$$2d^2 - a^2 = b^2 + c^2 + 2\sqrt{3}\,bc \sin \theta. \tag{4}$$

So, remembering that

$$\sin^2\theta + \cos^2\theta = 1,$$

we have from (2) and (4) that

$$(2d^2 - a^2 - b^2 - c^2)^2 + 3(-a^2 + b^2 + c^2)^2 = 12b^2c^2.$$

Tidying this up, we get both[3]

$$a^4 + b^4 + c^4 + d^4 = a^2b^2 + a^2c^2 + a^2d^2 + b^2c^2 + b^2d^2 + c^2d^2 \tag{5}$$

---

[3] Symmetrical in a, b, c, d. (Harry *did* say that the algebraical result was rather interesting!)

and

$$(2d^2 - a^2 - b^2 - c^2)^2 = 48\ \Delta^2, \tag{6}$$

where

$$\Delta = \tfrac{1}{4}\sqrt{\{(a+b+c)(-a+b+c)(a-b+c)(a+b-c)\}} \tag{7}$$

(Δ is the area of a triangle with sides a, b, c).

1.4   We still have a slight problem: (6) gives *two* values[4] for d^2 (it gives $d^2 = \tfrac{1}{2}(a^2 + b^2 + c^2) \pm 2\sqrt{3}\ \Delta$): which one should we take?[5]

One (at least) of $\widehat{YPZ}$, $\widehat{ZPX}$, $\widehat{XPY}$ is greater than or equal to $2\pi/3$; without loss of generality, we can assume it to be $\widehat{YPZ}$. Then

$$\begin{aligned} 2d^2 &\geqq 2(b^2 + bc + c^2)\\ &= a^2 + b^2 + c^2 + \{(b+c)^2 - a^2\}\\ &\geqq a^2 + b^2 + c^2. \end{aligned}$$

This resolves the ambiguity in (6). We now have unequivocally

$$d^2 = \tfrac{1}{2}(a^2 + b^2 + c^2) + 2\sqrt{3}\ \Delta, \tag{8}$$

where Δ is defined by (7).

This completes the *algebraic* part of the solution. There are, of course, other ways of arriving at (8): I discuss one of them—which uses neither trigonometry nor Anne's construction—in Appendix II.

2   Second part—the numbers

2.1 a, b, c, d are positive integers satisfying (1), (7), (8). d is the smallest integer for which this can be done. a, b, c are in arithmetic progression.

2.2   If a, b, c had a common prime factor (p, say), then by (5) we would have that p was a factor of d. We would then have that a/p, b/p, c/p, d/p were integers forming a valid solution, contrary to the datum that 'd is the smallest number for which this can be done.' So a, b, c have no common factor.

---

[4]   Dick: 'But Tom, there'd be *two* possible answers . . .'

[5]   And also: 'What is the significance of the other?' There is an alternative argument to that of para. 1.3: one that Tom had had in mind when he said 'It's as easy to go looking for both solutions in Anne's construction as it is to go looking for just one of them.' That alternative argument, which I give in Appendix I, explains the significance of the other value of d^2 given by (6)—but Tom may have been in error in thinking it as easy.

2.3   By symmetry, we can assume that b is the middle member of the Arithmetic Progression. Consequently there exists an integer k such that

$$a, c = b \pm k, \tag{9}$$

and (by para. 2.2) b, k have no common factor. By (1), (9) we have

$$5k < 2b, \tag{10}$$

and by (7), (8), (9),

$$2d^2 = 3b^2 + 2k^2 + 3b\sqrt{(b^2 - 4k^2)}. \tag{11}$$

So $b^2 - 4k^2$ must be a perfect square. So there must exist integers e, f (having no common factor) such that[6]

$$b = e^2 + f^2, \tag{12a}$$

$$k = ef. \tag{12b}$$

By symmetry, we can assume that

$$e > f. \tag{13}$$

By (11), (12), 13), we then have

$$d^2 = e^2(3e^2 + 4f^2). \tag{14}$$

So $3e^2 + 4f^2$ must be a perfect square. So there must exist integers p, q, r ($3q, r$ having no common factor) such that[7]

$$e = 2pqr, \tag{15a}$$

$$f = \tfrac{1}{2}p\left|3q^2 - r^2\right|, \tag{15b}$$

and p is 1 or 2 according as qr is odd or even.

By (13), (15) we have[8]

$$\sqrt{7} - 2 < r/q < \sqrt{7} + 2. \tag{16}$$

By (12), (14), (15) we have

$$b = \tfrac{1}{4}p^2(9q^2 + r^2)(q^2 + r^2), \tag{17a}$$

---

[6]   If $b^2 - 4k^2 = n^2$, then $(b+n)(b-n) = 4k^2$. So there exist integers e, f, g such that $b + n = 2ge^2$, $b - n = 2gf^2$, $k = gef$. So $b = g(e^2 + f^2)$. But b, k have no common factor. So $g = 1$, and e, f have no common factor.

[7]   By the same sort of reasoning as in the previous footnote, but a little more complicated. I leave it as an exercise for the reader. (Consider $(m + 2f)(m - 2f) = 3e^2$, and note that $3q^2 - r^2$ can not be divisible by 4 unless q, r are both even.)

[8]   We will tighten this inequality in a moment.

$$k = p^2qr|3q^2 - r^2|,$$ (17b)

$$d = 2p^2qr(3q^2 + r^2);$$ (17c)

and[9] by (10), (16), (17),

$$1 < r/q < 3.$$ (18)

2.4 The smallest value of d satisfying (17), (18) occurs when $q = 1$, $r = 2$ (and, so, $p = 2$): we then have

$$b = \ \ 65,$$
$$k = \ \ \ 8,$$
$$d = 112.$$

So the values of PX, PY, PZ that Tom had in mind are 57, 65, 73; and the side-length of the equilateral triangle is 112. (See Diagram 3.3.)

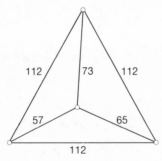

Diagram 3.3

Appendix I—Tom's alternative to para. 1.3

PQY is a triangle with sides a, b, c; write $\widehat{QPY} = \theta$. Then (see (2))

$$a^2 = b^2 + c^2 - 2bc \cos \theta.$$

On PQ, erect equilateral triangles PQZ, QPZ'; write $YZ = d$, $YZ' = d'$. (See Diagram 3.4, which includes Diagram 3.2.) Then

$$\left. \begin{array}{l} d^2 = b^2 + c^2 - 2bc \cos(\theta + \pi/3), \\ d'^2 = b^2 + c^2 - 2bc \cos(\theta - \pi/3). \end{array} \right\}$$

---

[9] From (10), (17) on their own, we have only that r/q must satisfy one of the three inequalities (i) $r/q < \sqrt{19} - 4$; (ii) $r/q > \sqrt{19} + 4$; and (iii) $1 < r/q < 3$. We need to observe (16) in order to eliminate the first two of these possibilities.

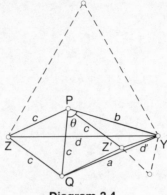

Diagram 3.4

Hence

$$d^2 + d'^2 = 2(b^2 + c^2) - 2bc\cos\theta,$$

and

$$d^2 d'^2 = b^4 - b^2 c^2 + c^4 - 2bc(b^2 + c^2)\cos\theta + 4b^2 c^2 \cos^2\theta.$$

So, by (2),

$$d^2 + d'^2 = a^2 + b^2 + c^2,$$

$$d^2 d'^2 = a^4 + b^4 + c^4 - (b^2 c^2 + c^2 a^2 + a^2 b^2).$$

Thus d^2, d'^2 are the roots of the equation

$$x^2 - (a^2 + b^2 + c^2)x + \{a^4 + b^4 + c^4 - (b^2 c^2 + c^2 a^2 + a^2 b^2)\} = 0,$$

which corresponds to (5).

This argument has an attractive symmetry, and is probably appropriate when one is looking for both of the sizes of equilateral triangle arising from 'Anne's construction'. But that is not our aim in the particular problem under discussion, for which the argument in the main text is both shorter and more straightforward.

Appendix II—No trigonometry

In this Appendix we obtain the algebraic solution (equations (7), (8) of the main text) without recourse to trigonometry. (The most advanced theorem used is Pythagoras.)

1   XYZ is an equilateral triangle and P is a point inside it. (See

Diagram 3.5.) We want to obtain d (the side-length of XYZ) in terms of a, b, c (the lengths of PX, PY, PZ, respectively).

2 Through P draw X_1X_2 parallel to YZ; Y_1Y_2 parallel to ZX; Z_1Z_2 parallel to XY (as in Diagram 3.6). Then

$$PY_1 = PZ_2 = Y_1Z_2 = X_1Y = X_2Z \ (=x, \text{ say})$$

$$PZ_1 = PX_2 = Z_1X_2 = Y_1Z = Y_2X \ (=y, \text{ say})$$

$$PX_1 = PY_2 = X_1Y_2 = Z_1X = Z_2Y \ (=z, \text{ say})$$

and so

$$d = x + y + z. \tag{19}$$

We also have that the perpendicular distance of P from ZX is $\frac{1}{2}\sqrt{3}\,y$, and that that perpendicular meets ZX at the mid-point of Z_1X_2, distant $z + \frac{1}{2}y$ from X; so, by Pythagoras,

$$a^2 = (\tfrac{1}{2}\sqrt{3}\,y)^2 + (z + \tfrac{1}{2}y)^2;$$

that is,

$$a^2 = y^2 + yz + z^2. \tag{20a}$$

Similarly,

$$b^2 = z^2 + zx + x^2, \tag{20b}$$

and

$$c^2 = x^2 + xy + y^2. \tag{20c}$$

3 Subtracting (20c) from (20b), we have

$$b^2 - c^2 = (z - y)(x + y + z),$$

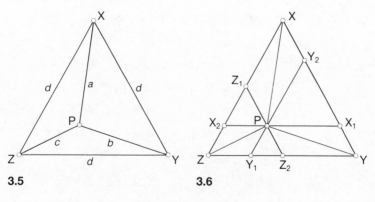

3.5   3.6

Diagrams 3.5–3.6

and so by (19)
$$z - y = (b^2 - c^2)/d. \tag{21}$$
Now
$$y^2 + 4yz + z^2 \equiv 2(y^2 + yz + z^2) - (z - y)^2,$$
so by (20a), (21)
$$y^2 + 4yz + z^2 = 2a^2 - (b^2 - c^2)^2/d^2. \tag{22a}$$
Similarly,
$$z^2 + 4zx + x^2 = 2b^2 - (c^2 - a^2)^2/d^2, \tag{22b}$$
and
$$x^2 + 4xy + y^2 = 2c^2 - (a^2 - b^2)^2/d^2. \tag{22c}$$

We now add up the three equations (22); the sum of the left-hand-sides is $2(x + y + z)^2$, so by (19)

$$d^2 = a^2 + b^2 + c^2 - (a^4 + b^4 + c^4 - b^2c^2 - c^2a^2 - a^2b^2)/d^2.$$

Multiplying through by d^2, we obtain (5)—and, so, (6), (7).

4　We still have to decide (as in para. 1.4) which of the two values of d^2 given by (6) provides the solution to our problem.

One (at least) of $\widehat{YPZ}$, $\widehat{ZPX}$, $\widehat{XPY}$ is greater than or equal to 120°; without loss of generality, we can assume it to be $\widehat{YPZ}$.

Drop a perpendicular from Z onto YP (extended), meeting it in R. (See Diagram 3.7.) By Pythagoras

$$YZ^2 = (YP + PR)^2 + RZ^2,$$
and
$$PZ^2 = PR^2 + RZ^2;$$
so (subtracting)
$$YZ^2 = PY^2 + 2PY.PR + PZ^2.$$

But $\widehat{RPZ} \leq 60°$, and so $2PR \geq PZ$. Hence

$$YZ^2 \geq PY^2 + PY.PZ + PZ^2,$$

which, as in para. 1.3, leads to (8).

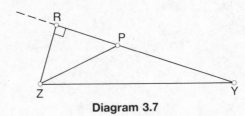

Diagram 3.7

Composer's problem

1 *Digression*

Chronologically, the three problems 'From The Point', 'To The Point', and 'Double Dealing' formed, for me, a most confusing trio. The algebraic solution of a 'From The Point'-type problem (with P *not* necessarily inside XYZ) came first; it suggested 'To The Point' (which had 'Double Dealing' as a spin-off); then back to the numeric solutions of 'From The Point'—which produced the idea of concentrating on the situation in which P was required to be inside XYZ; which changed the original algebraic approach.

So I had an underlying problem: in what order should the three problems be presented? (I am still not sure that the order that I have chosen is the most appropriate one—but it seemed to be the order in which Anne, Tom, Dick, and Harry were most likely to have met them.)

2 In the early stages of drafting 'From The Point', the two solutions of (6) were going to be given equal emphasis; hence Tom's suggested approach. It was only later that that symmetric approach was relegated to an Appendix.

3 The numeric part of the solution of 'From The Point' was—and still is—a trouble. When looking for integer solutions of (5), it is not too difficult[10] to find the solution 3, 5, 7, 8; but, whichever[11] one takes as the side-length of XYZ, one finds that P is outside XYZ (in three cases) or on its boundary (in the fourth).

Looking for other integer solutions of (5) (and in particular for one that would lead to P being strictly inside XYZ), I tried three separate approaches.

(i) Impose 'simplifying' constraints on a, b, c in (5) (or, rather, in (7)). I first tried the restraint '$(-a+b+c)$, $(a-b+c)$, $(a+b-c)$ must all be squares', and eventually obtained—I do not now recall how— the horrendous solution 3361, 40321, 42001, 43680. Well, at least it showed that a solution existed. But enough was enough.

(ii) Look for a general parametric solution. I got nowhere.

---

[10] It would have been less difficult (and would have saved a lot of paper) if I'd realized at the time that, if a, b, c, d are integers satisfying (5), then one of them must be divisible by 3, one of them must be divisible by 5, one of them must be divisible by 7, and one of them must be divisible by 8. But I didn't. (See 'Extension'.)

[11] Harry's 'Hang on; we can twiddle Anne's numbers round a bit' is an indirect reference to this.

(iii) Use a computer.[12] That was much more fruitful. Programmed to give, in ascending order of values of d, all strictly positive integer values of a, b, c, d that satisfied (1), (5), the printed output started with '57, 65, 73, 112'. Not only did I now have the smallest[13] integer solution of (1), (5), but the relationship between 57, 65, 73 sent me immediately back to (i) to try the 'simplifying constraint' 'a, b, c must be in arithmetic progression'. (I should, of course, have thought of that earlier. I didn't.) I now, as Anne recognized, had a numeric problem that could be solved without recourse to a computer.

Extension

1   Consider the integer equation (5):

$$a^4 + b^4 + c^4 + d^4 - a^2b^2 - a^2c^2 - a^2d^2 - b^2c^2 - b^2d^2 - c^2d^2 = 0.$$

If a, b, c, d had a common factor (p, say), we could divide through by p^4. So from now on let us assume that a, b, c, d have no common factor. It follows that no three of a, b, c, d have a common factor.

My first extension question is:

Can *two* of a, b, c, d have a common factor?

2   Looking in turn at the first few primes, we find that[14]

(i) Precisely one of a, b, c, d is divisible by 3;
(ii) Precisely one of a, b, c, d is divisible by 5;
(iii) Precisely one of a, b, c, d is divisible by 7;

---

[12] More accurately, I used a Professor of Computing Science. I gave a specification to Professor Mike Pitteway, of Brunel University, who wrote a program, ran it, and sent the output to me. (If one doesn't have a computer oneself, *always* have a Professor of Computing Science.)

[13] Computers are wonderfully useful in producing existence theorems. I could (and did) independently verify that 57, 65, 73, 112 satisfied (1), (5). But I still have not *independently* verified that 57, 65, 73, 112 is the smallest solution of (1), (5). I trust Mike's programming ability far more than I trust my own paper-and-pencil accuracy: but nevertheless I must do that paper-and-pencil verification—sometime. The point is a philosophical one, which Mike and I have debated on more than one occasion. (Mike is going to kill me for this footnote.)

[14] Any square on division by 3 leaves remainder 0 or 1; on division by 5 or by 8 leaves remainder 0, 1, or 4; and on division by 7 leaves remainder 0, 1, 2, or 4. It is readily verified that, if none of a, b, c, d were divisible by 3, or if just two of a, b, c, d were divisible by 3, then $a^4 + b^4 + c^4 + d^4 - a^2b^2 - a^2c^2 - a^2d^2 - b^2c^2 - b^2d^2 - c^2d^2$ would leave a non-zero remainder on division by 3, and so certainly could not be 0. Similarly with 5, 7, and 8.

(iv) Precisely one of a, b, c, d is divisible by 8—and each of the other three is odd.

Clearly, there are no other numbers of which we can say the same (since we know that there is a solution in which $a, b, c, d = 3, 5, 7, 8$).

But this comment is a side-issue that barely scratches the surface of the question in para. 1 above.

3   We have already (in para. 2.3 of the Solution) found a parametric solution of (5) that covers the particular case in which three of a, b, c, d are in Arithmetic Progression, namely

$$b = \tfrac{1}{4}p^2(9q^2 + r^2)(q^2 + r^2),$$

$$a, c = \tfrac{1}{4}p^2(9q^2 + r^2)(q^2 + r^2) \pm p^2 qr(3q^2 - r^2),$$

$$d = 2p^2 qr(3q^2 + r^2);$$

where $\mathrm{hcf}(3q, r) = 1$ and p is 1 or 2 according as qr is odd or even.

But that 'Arithmetic Progression' limitation is a very considerable one. So the main problem—my second and more difficult extension question—is:

> Can we find a parametric solution
> that covers the *general* case?

(*I* can't.)

4

WURSSFURSTIAN TRIELS

Problem I

Tom, Dick, and Harry went to Wurssfurstia for their summer holidays, and enthusiastically entered into the Wurssfurstian national sport—duelling. (Nobody gets hurt—they use paint-guns.)

The rules are simple: the duellist who is known to be the worse shot fires first; they fire alternately until one of them is hit; that one is deemed dead, and the other—the survivor—has won.

Since the results of all duels are published, any two potential duellists can work out their respective chances in their coming contest. (Apart from anything else, they need to do some calculation in order to decide who shall fire first.)

By the time that their Wurssfurstian holiday was over, Tom, Dick, and Harry each knew just what his chances were in a duel with either of the other two.

When they got home a friendly (well, not too unfriendly) dispute arose over which one of them was to be the first to take Anne out to dinner. To decide the point,[1] they held a Wurssfurstian Triel.[2]

They stood at the corners of an equilateral triangle, with the normal (Wurssfurstian) duelling distance between each pair of them. Tom was to fire first; Dick—known to be a better shot than Tom, but not as good as Harry—was (if not already 'dead') to fire second; Harry was (if not already 'dead') to fire third; and so on in the same cyclic order until there was a single survivor.

They all knew that in a duel between Tom and Dick the odds were 4:5 (in Dick's favour); that in a duel between Tom and Harry the odds

---

[1] In practice, of course, it was Anne who decided (and on a basis that had nothing to do with Wurssfurstian triels).

[2] A duel involves two people: a triel involves three people. (You can deduce the rules of a triel from the next paragraph.)

were 2:5 (in Harry's favour); and that in a duel between Dick and Harry the odds were 3:5 (in Harry's favour).

So: who was the favourite to win the triel—and what were his chances of doing so?

Problem II

Tom, Dick, and Harry had many Wurssfurstian duels after their first triel,[3] and when they eventually got together for a second triel their duelling relativities had changed significantly. Harry is still a better shot than Dick, who is still a better shot than Tom; but they all know that in a duel between Tom and Dick the odds are now 1:1, that in a duel between Tom and Harry the odds are now 3:5 (in Harry's favour), and that in a duel between Dick and Harry the odds are now 1:1.

Anne (who has, of course, worked things out in advance) is wondering whether one of them—one of them in particular—has realized just how different this triel should be from the previous one. (She needn't worry: he has.)

So: who is the favourite to win this second triel—and what are his chances of doing so?

[3] Described in Problem I. (The rules for the triel in this problem are the same as for the triel in Problem I.)

Problem III

Tom, Dick, and Harry are about to have their third—and last—triel.[4] Yet more practice—or lack of it—has changed their duelling relativities again: Harry is still a better shot than Dick, and Dick is still a better shot than Tom; but in a duel between Dick and Harry the odds are now 6:5 (in Dick's favour).

In this triel, each of the three has the same chance of being the ultimate survivor.[5]

So: in a duel between Tom and Dick the odds are . . . What? In whose favour?

---

[4] Their first two triels were described in Problems I and II; the rules for triels are given in Problem I.

[5] This is giving you rather more information than is necessary—all you really need to know is that a particular one of them has a 1-in-3 chance. But 'each of the three has the same chance' problems have a special interest (which I pursue in the Extension section).

General solution

1 *Bouts*

Before entering the complicated field of triels, or even the specialized one of duels, it is useful to discuss a simple form of contest, which I call a BOUT.

The rules for a bout are simple: there are two contestants; they stand the normal duelling distance apart; they fire alternately until one of them is hit; that one is deemed to be dead, and the other—the survivor—has won.

(Note that a *duel* is a special kind of *bout*—one in which there is the additional rule that 'the worse shot fires first'.)

Now let x be the probability that X will hit an aimed-at target (human size, at duelling distance) in one shot; and let y be the probability that Y will do so; and so on.

Suppose that X and Y have a large number (N, say) of bouts together, each time with X firing first. In Nx bouts, X hits Y with his first shot, and wins. In $N(1-x)y$ bouts, X misses Y with his first shot; but Y hits X with his first shot, and wins. (In the remaining $N(1-x)(1-y)$ bouts, each of X,Y misses with his first shot, and they effectively start again.) So, writing $R(X:Y)$ as the ratio of X's chance to Y's chance of winning a bout in which X fires first, we have

$$R(X:Y) = \frac{x}{(1-x)y}. \tag{1}$$

2 *Three duellists*

Three habitual duellists (A, B, C) know that A is a better shot than B and that B is a better shot than C; and they also know the values of the (possibly improper) fractions F, G, H, where

$$F = R(C:B), \quad G = R(C:A), \quad H = R(B:A). \tag{2}$$

By (1), (2) we then have

$$a = \frac{FH-G}{GH}, \quad b = \frac{FH-G}{FH}, \quad c = \frac{FH-G}{FH-G+H}. \tag{3}$$

(So from knowing the relative chances—in duels against each other—of A, B, C, we can deduce their individual abilities as marksmen.)

3 *The basic triel assumption*

A triel continues (by definition) until there is a single survivor. So in any particular triel,

> *either*   (i) it never ends;
> *and/or* (ii) someone, sometime, gets hit.

From now on I *assume* that someone, sometime, gets hit (and that the contestants make the same assumption).[6]

4 *The simplifying triel assumption*

I shall assume that the three contestants in a triel are of unequal ability[7] (A better than B, B better than C), and that each knows his chances in duels with each of the other two.

5 *The intelligent trielist*

While all three trielists are alive, each knows that, if he hits another, he will then be faced with a bout with the third (in which that third trielist will fire first). It is better for him to be in such a bout with the weaker of the other two than with the stronger. So (while all three are alive) each trielist, when it is his turn to fire, will either fire at the stronger of the other two or will delope (deliberately miss).

6 *A's reasoning*

A argues: 'Someone, sometime, gets hit [Section 3]. If either B or C makes the first hit, he'll hit *me* [Sections 4, 5]. So my chance of ultimate survival is zero unless *I* make the first hit. So I must not delope. So, when it comes to my turn to fire, I'll fire at B [Sections 4, 5]; when he's dead I'll turn my attention to C.'

7 *B's reasoning*[8]

B argues: 'While both A and C are still alive, A will be firing at me [Section 6], and C—if he's firing at anyone—will be firing at A

---

[6] This is key to A's reasoning (Section 6).

[7] If two (or three) of the trielists are of equal ability, the logic becomes Byzantine. (We could not, for example, form the conclusion stated in Section 5.)

[8] I give this in considerable detail. (I do not maintain this level of detail when it comes to a discussion of C's reasoning, which follows a similar pattern—though it does not necessarily come to the same conclusion.)

[Sections 4, 5]; I need to decide [Sections 4, 5] whether—when it comes to my turn to fire—to delope or to fire at A.

'If C adopts a "deloping" strategy, then my choice is fairly obvious: if I too were to delope, then A would eventually hit me—my chance of ultimate survival would be zero. So in that case I must not delope: I must fire at A [Sections 4, 5].

'But if C adopts a "fire at A" strategy, then my choice is not quite so obvious; I must do some sums.'

7.1 B's sums B needs to compare his chances of ultimate survival in the two cases:

(α) C fires at A, B delopes, A fires at B;
(β) C fires at A, B fires at A, A fires at B;

(during the period until one of them is hit).

Case (α)   In a large number (N, say) of triels,

> In Nc triels C hits A with his first shot; B's chance of ultimate survival is then his chance of winning a bout with C in which he (B) has first shot: namely,[9] $b/(b+c-bc)$.
>
> In $N(1-c)a$ triels, C (and B, by hypothesis) miss with their first shots, but A hits B with his: B's chance is zero.
>
> In the remaining $N(1-c)(1-a)$ triels, all three miss with their first shots, and effectively they start again.

So B's overall chance of ultimate survival is B_1^{**}, where

$$B_1^{**} = \frac{c \times \{b/(b+c-bc)\} + (1-c)a \times \{0\}}{1-(1-c)(1-a)} ;$$

that is,

$$B_1^{**} = \frac{bc}{(b+c-bc)(a+c-ac)} .$$

Case (β)   In a large number (N, say) of triels:

> In Nc triels C hits A with his first shot; B's chance of ultimate survival is then (as before) $b/(b+c-bc)$.
>
> In $N(1-c)b$ triels, C misses A with his first shot, but B hits A with his first shot; B's chance of ultimate survival is his chance of winning a bout with C in which C has first shot: namely $b(1-c)/(b+c-bc)$.

---

[9] By (1), B's chance, firing first against C, is $b/\{b+(1-b)c\}$.

In $N(1-c)(1-b)a$ triels, C and B miss with their first shots, but A hits B with his: B's chance is zero.

So B's overall chance of ultimate survival is B_1^*, where

$$B_1^* =$$

$$\frac{c \times \{b/(b+c-bc)\} + (1-c)b \times \{b(1-c)/(b+c-bc)\} + (1-c)(1-b)a \times \{0\}}{1-(1-c)(1-b)(1-a)};$$

that is,

$$B_1^* = \frac{b\{c+b(1-c)^2\}}{(b+c-bc)(a+b+c-bc-ca-ab+abc)}.$$

7.2 *B's strategy*   It is readily verified (since $1>a>b>c>0$) that B_1^* is greater than B_1^{**}. So if C adopts a 'fire at A' strategy, B's best strategy is to do the same. But this is the same strategy (for B) as the one that he has already decided to adopt if C adopts a 'deloping' strategy. So, having done his sums, B decides: 'Irrespective of what C decides to do, when it comes to my turn to fire, I'll fire at A; when he's dead I'll turn my attention to C.'

8   C's reasoning

C argues: 'While both A and B are still alive, they'll fire at each other, ignoring me [Sections 6, 7]; only when one of them is dead will the survivor turn his attention to me. In the meantime: should I delope (in which case I'll at least have first shot at the survivor—though that could well be A), or should I try to hit A (when if I succeed I know that it will only be B that I'll be up against—but he'll have first shot at me)? I must do some sums.'

8.1 *C's sums*   C needs to compare his chances of ultimate survival in the two cases:[10]

    (γ)   C delopes, B fires at A, A fires at B;
    (β)   C fires at A, B fires at A, A fires at B;

(during the period until one of them is hit).

    Although C is naturally most concerned with his own chances, we would like him also to calculate A's and B's chances under each of the two possible strategies that he (C) might adopt. So he does.

---

[10] The second case is the same as the second case that B considered.

Following the same sort of reasoning that B used (para. 7.1), we obtain:

Case (γ) A's, B's, and C's chances of ultimate survival are respectively A_0^*, B_0^*, C_0^*, where[11]

$$A_0^* = \frac{a^2(1-b)(1-c)}{(c+a-ca)(a+b-ab)}, \tag{4a}$$

$$B_0^* = \frac{b^2(1-c)}{(b+c-bc)(a+b-ab)}, \tag{4b}$$

$$C_0^* = 1 - A_0^* - B_0^*. \tag{4c}$$

Case (β) A's, B's, and C's chances of ultimate survival are respectively A_1^*, B_1^*, C_1^*, where

$$A_1^* = \frac{a^2(1-b)(1-c)^2}{(c+a-ca)(a+b+c-bc-ca-ab+abc)}, \tag{5a}$$

$$B_1^* = \frac{b\{c+b(1-c)^2\}}{(b+c-bc)(a+b+c-bc-ca-ab+abc)}, \tag{5b}$$

$$C_1^* = 1 - A_1^* - B_1^*. \tag{5c}$$

8.2 *C's strategy* After some rather tedious but very elementary algebra, we have from (4), (5) that

$$C_0^* - C_1^* = \frac{c^2[\{b^2(1-a)+a^2(1-b)^2\}c - a\{a(1-b)^2-b\}]}{(b+c-bc)(c+a-ca)(a+b-ab)(a+b+c-bc-ca-ab+abc)}. \tag{6}$$

So C selects strategy γ ('Delope') if $c > \Gamma(a,b)$, and strategy β ('Fire at A') if $c < \Gamma(a,b)$, where

$$\Gamma(a,b) = \frac{a\{a(1-b)^2-b\}}{b^2(1-a)+a^2(1-b)^2}. \tag{7}$$

8.3 *Notes on (6)*

8.31 It follows from (6) that, if C is a fairly good shot, he should

---

[11] The formula for C_0^* in terms of a, b, c is a little lengthy. Writing it as in (4c) is shorter, emphasizes that the three chances sum to unity, and helps (a little) in para. 8.2.

adopt strategy γ (Delope)—until one of A and B has hit the other.

8.32   Irrespective of how poor a shot he is, C should always adopt strategy γ if $b \geq a(1-b)^2$.

8.33   In particular, C should always adopt strategy γ if $R(B:A) \geq 1$. For then, by (1), $b \geq a(1-b)$, and so *a fortiori* $b > a(1-b)^2$.

9   *Alternative notation*

It is sometimes useful to write (4), (5), (6), (7) in terms of F, G, H rather than in terms of a, b, c.

9.1   *Case γ*   By (3) we have from (4) that, when C adopts the strategy of deloping (until A or B is dead), then

$$\left.\begin{aligned} A_0^* &= \frac{1}{(G+1)(H+1)}, \\[2mm] B_0^* &= \frac{H}{(F+1)(H+1)}, \\[2mm] C_0^* &= 1 - A_0^* - B_0. \end{aligned}\right\} \tag{8}$$

9.2   *Case β*   By (3) we have from (5) that, when C adopts the strategy of firing at A (until he is dead), then

$$\left.\begin{aligned} A_1^* &= \frac{1}{(G+1)(FH+H+1)}, \\[2mm] B_1^* &= \frac{H(F^2+F+1)-FG}{(F+1)(FH+H+1)}, \\[2mm] C_1^* &= 1 - A_1^* - B_1^*. \end{aligned}\right\} \tag{9}$$

9.3   *The sign of $C_0^* - C_1^*$*   By (3) we have[12] from (6) that

$$C_0^* - C_1^* = \frac{F\{FH(GH+G+H)-G(GH+G+1)\}}{(F+1)(G+1)(H+1)(FH+H+1)}. \tag{10}$$

---

[12] It would have considerably reduced the tediousness of the algebra in para. 8.2 if we had developed (8) and (9) *before* calculating $C_0^* - C_1^*$, and then obtained (10) directly from (8), (9). But in that case we would not—yet—have had (6); and so we would not have had para. 8.3. (I do not think that it is obvious from (10) that 'the better a shot C is, the more it is in his interest to delope'; and that interesting result *is* obvious from (6).)

So C selects strategy γ ('Delope') if $F > \Phi(G, H)$, and strategy β ('Fire at A') if $F < \Phi(G, H)$, where

$$\Phi(G, H) = \frac{G(GH + G + 1)}{H(GH + G + H)}. \tag{11}$$

Particular solution I

In the notation of the General Solution, we have

$$\text{Tom} \equiv C, \quad \text{Dick} \equiv B, \quad \text{Harry} \equiv A;$$

and

$$F(\equiv R(C:B)) = 4/5,$$

$$G(\equiv R(C:A)) = 2/5,$$

$$H(\equiv R(B:A)) = 3/5.$$

So by (11) we have that $\Phi(G, H) = 82/93$, which is greater than F. Consequently (by para. 9.3), C adopts strategy β (that is, whenever it is his turn to fire, he tries to hit A—until A is dead). Hence, by (9), it follows that the chances of the trielists are:

$$A_1^* = 1125/3276,$$

$$B_1^* = 1001/3276,$$

$$C_1^* = 1150/3276.$$

So Tom (C) is the favourite, and his chances are 575/1638.

We could, alternatively, have calculated a, b, c from (3) (even if only to check that the given values of F, G, H really did imply that $1 > a > b > c > 0$), getting $a = 1/3$, $b = 1/6$, $c = 2/17$; then calculated $\Gamma(a,b) = 7/31$, which is greater than c; and so gone to (5) to calculate A_1^*, B_1^*, C_1^*.

(Incidentally, if Tom had erroneously adopted strategy γ (that is, if he'd decided to delope until one of A, B was dead), the chances of the trielists would have been, by (8),

$$A_0^* = 75/168,$$

$$B_0^* = 35/168,$$

$$C_0^* = 58/168;$$

and A would have had the best chance of winning.)

Particular solution II

In the notation of the General Solution, we have

$$\text{Tom} \equiv C, \quad \text{Dick} \equiv B, \quad \text{Harry} \equiv A;$$

and

$$F(\equiv R(C\!:\!B)) = 1,$$

$$G(\equiv R(C\!:\!A)) = 3/5,$$

$$H(\equiv R(B\!:\!A)) = 1.$$

We know (by para. 8.33) that C should adopt strategy γ (that is, whenever it is his turn to fire, he delopes—until one of A, B is dead). Hence, by (8), it follows that the chances of the trielists are:

$$A_0^* = 5/16,$$

$$B_0^* = 4/16,$$

$$C_0^* = 7/16.$$

So Tom (C) is the favourite, and his chances are 7/16.

We could, alternatively, have calculated a, b, c from (3), getting $a = 2/3$, $b = 2/5$, $c = 2/7$;[13] and then (because of para. 8.33) gone straight to (4) to calculate A_0^*, B_0^*, C_0^*.

(Incidentally, if Tom had erroneously adopted strategy β (that is, if he'd decided to try to hit A even though B was still alive),[14] the chances of the trielists would have been, (9),

$$A_1^* = 25/120,$$

$$B_1^* = 48/120,$$

$$C_1^* = 47/120;$$

and B would have had the best chance of winning.)

Particular solution III

In the notation of the General Solution, we have

$$\text{Tom} \equiv C, \quad \text{Dick} \equiv B, \quad \text{Harry} \equiv A;$$

---

[13] They *have* improved, haven't they?
[14] This is what Anne was wondering about: whether Tom had realized that deloping (in the early stages) was a viable strategy—and that it was the correct one to adopt in this particular triel.

and

$$H(\equiv R(B:A)) = 6/5.$$

Since $H > 1$, we have (by para. 8.33) that C should adopt strategy γ. Consequently, by (8) (and using $H = 6/5$), the chances of the trielists are:

$$A_0^* = 5/11(G+1),$$
$$B_0^* = 6/11(F+1),$$
$$C_0^* = 1 - A_0^* - B_0^*.$$

But we have been told that these are equal. Consequently

$$F = 7/11,$$
$$G = 4/11.$$

So in a duel between Tom (C) and Dick (B), the odds are 7:11, in Dick's favour.

For completeness, we calculate a, b, c from (3) (check that the situation is a possible one), and get $a = 11/12$, $b = 11/21$, $c = 1/4$.[15]

Composer's problem

1   *How to get them started*

One trouble in the setting of this problem lay in getting the contestants to fight at all. Let's suppose for a moment that A is a very very good shot, that B is a very good shot, and that C is a good shot. A and B know that while they're both still alive C will delope, but that as soon as one of them hits the other the survivor will very probably be killed by C (who will have the first shot in the ensuing bout). So there is a strong temptation for A and B to agree *not* to hit each other. The agreement may well not be overt: B, when it is his turn to fire, may ostentatiously delope (ignoring Section 7 of the General Solution!), implicitly suggesting to A that he should do the same. If A is confident that B will continue to delope, then it is to A's advantage to delope also. The triel will become a succession of deliberate misses—with no losers, and, so, no ultimate single survivor. (The three will probably

---

[15] Dick and Harry have improved still further; but Tom, I'm afraid, has dropped off a little. Still, he's a lot better than he was to begin with.

finish up in the Ayling Arms grumbling about the cost of the wasted ammunition.)

It is because of this difficulty that I introduced 'nobody gets hurt—they use paint-guns'. (Tom, Dick, and Harry are intelligent; and, if they'd had to use real guns, the intelligent thing to do would have been to drop the whole idea of a triel and to go straight to the Ayling Arms in the first place.) It was also my reason for '. . . and so on . . . until there was a single survivor': I needed it for Section 3 of the General Solution; the assumption there is key to A's reasoning in Section 6 of the General Solution.

2   *The order of the three problems*

Originally I had intended to give just Problem III.[16] But that problem leads to a horrendous amount of (incorrect) work if the solver has not, fairly early on, realized the possibility of deloping. So I introduced Problems I and II as pointers. In Problem I, Tom (C) should not delope, and so solvers will still get the right answer even if they haven't thought of deloping (provided, of course, that they realize that A and B will fire at each other until one of them is dead). Such a solver will not, however, get the right answer to Problem II; I hope that, alerted by Anne's concerns, he will find out the deloping option before tackling Problem III.

3   *The missing problem IV*

Problem III is an 'Equal Opportunity' problem in which C (correctly) adopts the Delope option. I would have liked to have had a corresponding Equal Opportunity problem in which C (correctly) adopts the 'Fire at A' option. But I could not find a (non-trivial) *rational* example. (At least this has provided me with a good Extension Question—but, all the same, it's been most frustrating!)

4   *The missing problem V*

Going on from there, I would have liked to have set a problem giving as data the value of *F* and the constraint of Equal Opportunity—a problem that had *two* answers.

---

[16] First published, in a modified form, in *IBM UK News* (19 December 1986).

EXAMPLE: Given that $F = 3/5$ and that each of the trielists has the same chance of being the ultimate survivor, there are *two* possible answers: one with $G = 2/5$ and $H = 8/7$, the other with $G = (\sqrt{8593} - 73)/48$ and $H = (15\sqrt{8593} - 455)/1328$. (I leave it to you to verify that C is correct to delope in the first case and to fire at A in the second—and to verify that in both cases $1 > a > b > c > 0$.)

But, if I cannot find a rational example for Problem IV, I certainly cannot find one for Problem V. (And I am not going to set either problem until I do!)

Extension
(LET'S BE RATIONAL ABOUT EQUAL OPPORTUNITY)

1   *Equal opportunity*

In this extension I confine myself to situations in which each of the trielists has the same chance of being the ultimate survivor—that is, to situations in which

$$A^* = B^* = C^* = 1/3 \tag{12}$$

(where (12) is either $A_0^* = B_0^* = C_0^* = 1/3$ or $A_1^* = B_1^* = C_1^* = 1/3$; we need to develop a test to determine which).

2   *The test*

If $H > 1$, we have (by para. 8.33) that C is correct to adopt strategy γ (Delope).

   In Equal Opportunity situations, this has a simple converse. We argue:

   If C is correct to adopt strategy γ, then $A_0^* = B_0^* = C_0^* = 1/3$ (by (12)), and so by (8),

$$(G + 1)(H + 1) = 3,$$

$$(F + 1)(H + 1) = 3H.$$

Then by (3)

$$\frac{a}{b} = \frac{F}{G} = \frac{2H - 1}{2 - H}.$$

But $a > b$, and so $H > 1$.

So if C is correct to adopt strategy γ, then $H > 1$.[17]

We can summarize this as:[18]

If $H > 1$: then (12) is $A_0^* = B_0^* = C_0^* = \frac{1}{3}$,
If $H < 1$: then (12) is $A_1^* = B_1^* = C_1^* = \frac{1}{3}$.

3 $H > 1$

If $H > 1$, we have (3), (8), and

$$A_0^* = B_0^* = C_0^* = \tfrac{1}{3},$$

from which we have virtually immediately the complete parametric solution

$$F = 1 - G, \quad H = \frac{2 - G}{1 + G};$$

$$a = \frac{2(1 - 2G)}{G(2 - G)}, \quad b = \frac{2(1 - 2G)}{(1 - G)(2 - G)}, \quad c = \frac{2(1 - 2G)}{4 - 5G};$$

subject only to the limits[19]

$$3 - \sqrt{7} \leqq G < \tfrac{1}{2}.$$

If we insert any rational value of G (between these limits) we have an Equal Opportunity problem in which all the relevant quantities are also rational (and in which C correctly chooses strategy γ—Delope).

4 $H < 1$

If $H < 1$, we have (3), (9), and

$$A_1^* = B_1^* = C_1^* = \tfrac{1}{3},$$

from which we can obtain a complete parametric solution, the first part of which is

$$F = \frac{G(2 + G) - \sqrt{(G^4 + 12G - 4)}}{2(1 - 2G - G^2)},$$

---

[17] It should perhaps be emphasized that this statement is made in the context of Equal Opportunity. It is not valid generally.

[18] '$H = 1$' is impossible, since (with (12)) it implies $a = b = c = 0$.

[19] The limits are, basically, imposed by the requirement $1 \geqq a > 0$; all other requirements are then met.

$$H = \frac{2 - 2G - G^2 + \sqrt{(G^4 + 12G - 4)}}{2(1+G)};$$

subject only to the limits[20]

$$\frac{\sqrt{10} - 2}{3} \leqq G < \sqrt{2} - 1.$$

If we can find a rational value of G (between these limits) such that $\sqrt{(G^4 + 12G - 4)}$ is also rational, then we will have an Equal Opportunity problem in which all the relevant quantities are also rational (and in which C correctly chooses strategy β—Fire at A).

5   Extension question

The equation

$$G^4 + 12G - 4 = N^2 \tag{13}$$

certainly has rational solutions[21] (e.g. $G = 1/3$, $N = 1/9$; $G = 1$, $N = 3$; $G = 2$, $N = 6$); but none of those that I have found has met the requirement

$$\frac{\sqrt{10} - 2}{3} \leqq G < \sqrt{2} - 1. \tag{14}$$

So my Extension Question is: Can you find rational numbers G, N satisfying (13), (14)?

[20] The upper limit is imposed by the requirement $a > 0$; the lower limit by the requirement $F \leqq \Phi$; all other requirements are then met. (Had this been part of the General Solution, rather than part of the Extension, then this assertion would have occupied a page of text, rather than two lines of a footnote! The justifying algebra is left as an enjoyment for the reader.)

[21] And, so, an infinity of rational solutions.

5

STOP WATCH

Problem I[1]

Tom's watch is pretty accurate—when it's working. It has a sweep second-hand that advances in jerks, one each second: if the watch stops, it stops with the second-hand pointing directly to one of the 'second' (or 'minute') marks round the rim. At noon all three hands are (naturally) pointing directly at the zero-minute mark.

At noon yesterday Tom's watch was certainly working, but when he checked it at 6.00 p.m. he found that it had stopped some time before. The angles the three hands made with each other all differed from 120° by less than 1°.

At what time (hours, minutes, seconds) had Tom's watch stopped?

---

[1] First published (in slightly modified form) on 30 October 1986 as Enigma 383 in the *New Scientist*.

Problem II

Dick's watch differs from Tom's only in that its sweep second-hand advances in half-second jerks (instead of in whole-second ones); so it can stop with the second-hand in any one of 120 positions (instead of just 60).

'Suppose,' said Dick, 'that the same sort of thing[2] had happened to me—that my watch had stopped at some time between noon and 6.00 p.m. and that the angles that the three hands made with each other all differed from 120° by less than $\frac{1}{2}°$. What time would it have stopped at?'

---

[2] The same sort of thing that happened to Tom in Problem I.

Problem III

Harry's watch has a sweep second-hand that sweeps smoothly—no jerks at all: it can stop with the second-hand pointing anywhere (that is, with no restrictions about whole numbers of seconds, as in Tom's case, or whole numbers of half-seconds, as in Dick's case).

'If the same sort of thing[3] had happened to *me*,' said Harry, 'my watch would have stopped some time between noon and 6.00 p.m. with the angles the three hands made with each other all differing from 120° by no more than iota degrees. What time would it have stopped at?'

'Iota?' said Anne. 'What's iota?'

'Iota is the smallest number for which the situation is possible at all,' said Harry; 'so I'm asking you to tell me what iota is—as well as what time my watch would have stopped at.'

'Well, obviously iota's less than a half,' said Anne, 'but I think that it's going to turn out to be a rather nasty fraction. Would a close approximation do? Or does it have to be exact?'

'It has to be exact,' said Harry, '—and that goes for the time, too.'

So: what is iota? And at what time would Harry's watch have stopped?

---

[3] The same sort of thing that happened to Tom in Problem I and to Dick in Problem II.

General solution

1

The three hands, H, M, S (Hour, Minute, Second) of a clock (or watch) are perfectly aligned at noon.

At τ hours[4] after noon H, M, S have moved (clockwise) through $30\tau°$, $360\tau°$, $21600\tau°$ respectively.

2   *Congruences*

A hand that is in a position (measured clockwise from the noon position) $\theta°$ $(0 \leqq \theta < 360)$ may have reached that position simply by moving through $\theta°$. But it may also have reached that position by moving through $(360g + \theta)°$, where g is any integer.

So to the *position* $\theta°$ $(0 \leqq \theta < 360)$ there correspond a multitude of *position-descriptions* $(360g + \theta)°$, where g takes all integer values. I refer to these different descriptions of the same position as being *congruent* to each other, and write

$$(360f + \theta)° \equiv (360g + \theta)°$$

for all integer f, g.

Conversely: if we know that a hand has moved (clockwise from the noon position) through $\phi°$, then we know that its position is $(\phi - 360h)°$ for some integer value of h (specifically: for that integer value of h that makes $0 \leqq \phi - 360h < 360$). When—but not until—we know ϕ explicitly, we can calculate the integer h, and so obtain the hand's position.

> EXAMPLE: At time 1 hr., 23 min., 45 sec. (i.e. at $1\frac{19}{48}$ hours after noon) H, M, S have moved through $41\frac{7}{8}°$, $502\frac{1}{2}°$, $30150°$ respectively. These angles are *position-descriptions* of the hands. But $502\frac{1}{2}° = (360 + 142\frac{1}{2})° \equiv 142\frac{1}{2}°$, and $30150° = ((360 \times 83) + 270)° \equiv 270°$. So the *positions* of the hands are $41\frac{7}{8}°$, $142\frac{1}{2}°$, $270°$.

3   *Reflections*

$(12 - \tau)$ hours after noon H, M, S have moved through $(360 - 30\tau)°$, $(4320 - 360\tau)°$, $(259200 - 21600\tau)°$ respectively. So (taking con-

---

[4] I use lower-case *italic* letters to denote integers; I use Greek letters to denote quantities that are not necessarily integers.

gruences) their positions (measured clockwise from the noon position) can be described as $-30\tau°$, $-360\tau°$, $-21600\tau°$. So H, M, S are in positions that are reflections (in the line through the noon position) of their positions τ hours after noon.

> EXAMPLE: At time 10 hr., 36 min., 15 sec. (i.e. at $12 - 1\frac{19}{48}$ hours after noon) H, M, S have moved through $318\frac{1}{8}°$, $3817\frac{1}{2}°$, $229050°$ respectively. Taking congruences, their positions are $318\frac{1}{8}°(\equiv -41\frac{7}{8}°)$, $217\frac{1}{2}°(\equiv -142\frac{1}{2}°)$, $90°(\equiv -270°)$: these positions are the reflections of the positions of the hands at time 1 hr., 23 min., 45 sec. (i.e. at $1\frac{19}{48}$ hours after noon). These reflective situations are illustrated in Diagram 5.1.

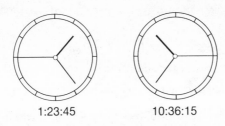

1:23:45   10:36:15

Diagram 5.1

4 *Hand order*

One consequence of Section 3 is that we can confine our attention (at least until towards the end of our analysis) to situations in which the clockwise cyclic order of the hands round the clock is H→M→S (since any configuration in which the clockwise cyclic order is H→S→M must be the reflection of a configuration in which the cyclic order is H→M→S).[5] So we make that assumption throughout Section 5.

5 $\widehat{HOM}, \widehat{MOS}, \widehat{SOH} \cong 120°$

Let us define α, β, γ by

$$\widehat{HOM} = (120 + \alpha)°, \qquad \widehat{MOS} = (120 + \beta)°, \qquad \widehat{SOH} = (120 + \gamma)°;$$

and confine our attention to situations in which, for some reasonably

---

[5] We must not, of course, *also* confine our attention to situations in which the time is between noon and 6.00 p.m. (We can do one or the other, but not both together.)

small[6] number δ,

$$|\alpha|, |\beta|, |\gamma| < \delta.$$

We note that

$$\alpha + \beta + \gamma = 0.$$

Such a situation occurs at τ ($0 \leq \tau < 12$) hours after noon (see Section 1) if there are integers m, n such that

$$(360 - 30)\tau = 360m + 120 + \alpha \quad (0 \leq m \leq 10), \tag{1}$$

$$(21600 - 360)\tau = 360n + 120 + \beta \quad (0 \leq n \leq 707), \tag{2}$$

$$|\alpha|, |\beta|, |\alpha + \beta| < \delta. \tag{3}$$

Eliminating τ between (1) and (2), we have

$$360(708m - 11n + 232) = 11\beta - 708\alpha - 120. \tag{4}$$

Since m, n are integers, it follows from (4) that there is an integer k such that

$$708m - 11n + 232 = k, \tag{5}$$

and

$$11\beta - 708\alpha = 120(3k + 1). \tag{6}$$

Expressions (3) and (6) provide us with limits for the possible (let us call them 'acceptable') values of k: k is an integer such that

$$\frac{-719\delta - 120}{360} \leq k \leq \frac{719\delta - 120}{360}. \tag{7}$$

EXAMPLE: When we take $\delta = 1$, (7) tells us that the acceptable values of k lie in $-\frac{839}{360} \leq k \leq \frac{599}{360}$, so that (since k is an integer) k must be one of -2, -1, 0, 1.

For each of the acceptable values of k we then solve the integer equation (5),[7] to obtain the corresponding 'acceptable' values of m

---

[6] What follows is valid for any $\delta \leq 60$. But it is probably better to think of δ as being appreciably smaller than that. (In the three Stop Watch problems $\delta \leq 1$.)

[7] In practice we don't 'solve' (5) in this way at all: what we do is to give m the eleven values $0, 1, \ldots, 10$ (in turn), and then see, for each of them, what values of k make $708m + 232 - k$ divisible by 11.

and n; getting Table 5.1[8] (or, rather, a part of it: the *numbers* in the table are the same in all cases; it is the relevant *extent* of the table that is determined by the value of δ, by (7)).

Table 5.1

k	m	n
$\dots -16, -5, \quad 6, \dots$	4	$\dots 280, 279, 278, \dots$
$\dots -15, -4, \quad 7, \dots$	7	$\dots 473, 472, 471, \dots$
$\dots -14, -3, \quad 8, \dots$	10	$\dots 666, 665, 664, \dots$
$\dots -13, -2, \quad 9, \dots$	2	$\dots 151, 150, 149, \dots$
$\dots -12, -1, 10, \dots$	5	$\dots 344, 343, 342, \dots$
$\dots -11, \quad 0, 11, \dots$	8	$\dots 537, 536, 535, \dots$
$\dots -10, \quad 1, 12, \dots$	0	$\dots \quad 22, \quad 21, \quad 20, \dots$
$\dots \quad -9, \quad 2, 13, \dots$	3	$\dots 215, 214, 213, \dots$
$\dots \quad -8, \quad 3, 14, \dots$	6	$\dots 408, 407, 406, \dots$
$\dots \quad -7, \quad 4, 15, \dots$	9	$\dots 601, 600, 599, \dots$
$\dots \quad -6, \quad 5, 16, \dots$	1	$\dots \quad 86, \quad 85, \quad 84, \dots$

EXAMPLE: Continuing with the case of $\delta = 1$, the relevant part of Table 5.1 is Table 5.2.

Table 5.2

k	m	n
-2	2	150
-1	5	343
0	8	536
1	0	21

For any specific value of δ, we can now work out,[9] using (1), the ranges of times for which (3) is satisfied.

In doing so it may be helpful first to sketch, in the (α, β)-plane, both (3) (which is δ-dependent) and (6) (which is not δ-dependent). The sketch of (3) is a hexagonal region. The sketch of (6) consists of a number of parallel lines, one for each value of k, with gradient $\frac{708}{11}$; those for acceptable value of k intersect the region (3), This helps in determining the two limiting values of α to insert in (1) for each of the acceptable values of m.

---

[8] Table 5.1 is included only for the sake of completeness of the argument. For all practical purposes (and certainly for the three 'Stop Watch' problems) we need only a small subset of it—as in the example.

[9] The arithmetic is a little tedious, but straightforward.

EXAMPLE: Continuing with the case of $\delta = 1$, we first make the sketch suggested above (Diagram 5.2). For each of the pairs k, m in Table 5.2, we calculate from (3), (6) (keeping an eye on Diagram 5.2) the ranges of times for which (3) is satisfied (Table 5.3).

If the clock's second-hand advances in jerks, rather than continuously, then when the clock stops it does so with the second hand a whole number of jerks away from the noon position. We are interested in those positions (if any) in which the clock can stop that fall in any of the ranges.

Table 5.3

k	m	Limits of α		$\|\alpha\|, \|\beta\|, \|\alpha+\beta\| \leqq 1$		
		from	to	at time	from	to
-2	2	$\frac{589}{708}$	$\frac{611}{719}$	2 hr., 32 min., $+\ 52\frac{42}{59}$ sec.		$52\frac{652}{719}$ sec.
-1	5	$\frac{229}{708}$	$\frac{251}{719}$	5 hr., 49 min., $+\ \ 8\frac{58}{59}$ sec.		$9\frac{189}{719}$ sec.
0	8	$-\frac{131}{719}$	$-\frac{109}{708}$	9 hr., 5 min., $+\ 25\frac{205}{719}$ sec.		$25\frac{35}{59}$ sec.
1	0	$-\frac{491}{719}$	$-\frac{469}{708}$	0 hr., 21 min., $+\ 41\frac{461}{719}$ sec.		$41\frac{51}{59}$ sec.

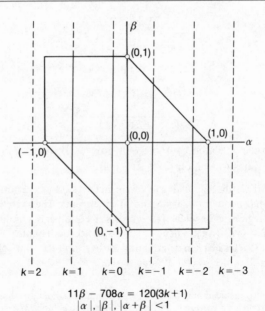

$$11\beta - 708\alpha = 120(3k+1)$$
$$\|\alpha\|, \|\beta\|, \|\alpha+\beta\| < 1$$

Diagram 5.2

EXAMPLE: Continuing with the case of $\delta = 1$, let us also suppose that the clock can stop only after a whole number of seconds. Examination of the ranges that we have developed shows that there is just one possible time for which (3) is satisfied:

5 hr., 49 min., 9 sec.

(As a check, we calculate the positions of the hands at this time: H @ $174\frac{23}{40}°$, M @ $294\frac{9}{10}°$, S @ $54°$; so that $\alpha = \frac{13}{40}$, $\beta = -\frac{9}{10}$, $\gamma = \frac{23}{40}$, verifying that $|\alpha|$, $|\beta|$, $|\gamma|$ are all less than δ ($=1$).)

6   *The other hand order*

Throughout Section 5 we have assumed that the clockwise cyclic order of the hands round the clock is $H \rightarrow M \rightarrow S$. We now need to add to our list the positions for which the clockwise cyclic order of the hands round the clock is $H \rightarrow S \rightarrow M$. These are, simply, the reflections (in the line through the noon position) of the $H \rightarrow M \rightarrow S$ position(s) that we have already obtained. (See Sections 3, 4.)

EXAMPLE: Continuing with the case of $\delta = 1$, and a clock that can stop only after a whole number of seconds, we have that there are just two times that meet the conditions that we have—so far—imposed:

5 hr., 49 min.,   9 sec.,
6 hr., 10 min., 51 sec.

7   *The last restraint*

Finally, we need to take account of any further restraint imposed in the statement of the problem. (Typically, it will be a restraint that rules out all but one of the solutions that we have so far obtained.)

EXAMPLE: Continuing with the case of $\delta = 1$, and a clock that can stop only after a whole number of seconds, if the additional restraint is imposed that (say) the time must be between noon and 6.00 p.m., then there is only one possible solution:

5 hr., 40 min., 9 sec.

Particular solution I

Tom's watch stops after a whole number of seconds; when it stops, the angles between the hands all differ from $120°$ by less than $1°$; and the time is between noon and 6.00 p.m.

This is the problem that we have been using as an example in Sections 5, 6, 7 of the General Solution, from which we have that there is a unique solution:

5 hr., 49 min., 9 sec.

Particular solution II

Dick's watch stops after a whole number of half-seconds; when it stops, the angles between the hands all differ from 120° by less than $\frac{1}{2}°$; and the time is between noon and 6.00 p.m.

There are two ways in which we can tackle this problem (and which one one chooses is a matter of individual taste). In one method we use *all* the results obtained in the General Solution (including the worked example—the solution of Problem I); in the other method we use only the *procedure* of the General Solution.

Method I

From Table 5.3, we have that there is no possible result for $k = -2$ or for $k = 1$ (since no half-second point then meets the requirement $|\alpha|$, $|\beta|$, $|\alpha+\beta| \leq 1$, never mind our—now—tighter requirement $|\alpha|$, $|\beta|$, $|\alpha+\beta| \leq \frac{1}{2}$).

If there is a result for $k = -1$, then it must be at 5 hr., 49 min., 9 sec.; and from the check that we made at the end of the worked example, we know that here $|\beta| = \frac{9}{10} > \frac{1}{2}$.

So—if there is a result at all—it must occur when $k = 0$, and must be 9 hr., 5 min., $25\frac{1}{2}$ sec. (or its reflection).

Since the time is between noon and 6.00 p.m., the only possibility is

2 hr., 54 min., $34\frac{1}{2}$ sec.

We calculate the positions of the hands at this time: H @ $87\frac{23}{80}°$, M @ $327\frac{9}{20}°$, S @ 207°; so that $\alpha = -\frac{23}{80}$, $\beta = \frac{9}{20}$, $\gamma = -\frac{13}{80}$. So at this time all of $|\alpha|$, $|\beta|$, $|\gamma|$ are less than $\frac{1}{2}$. So our 'only possibility' is a valid result.

Method II

We follow the procedure of the General Solution, with $\delta = \frac{1}{2}$.

(7) then tells us that the acceptable values of k lie in $-\frac{959}{720} \leq k \leq \frac{479}{720}$, so that k must be one of -1, 0.

So the relevant part of Table 5.1 is Table 5.4.

Table 5.4

k	m	n
1	5	343
0	8	536

We now sketch (3) (with $\delta = \frac{1}{2}$) and (6) in Diagram 5.3.

For each of the pairs k, m in Table 5.4, we calculate from (3), (6) (keeping an eye on Diagram 5.3) the limiting values of α; and we then calculate from (1) the ranges of times for which (3) is satisfied (Table 5.5).

Since Dick's watch stops after a whole number of half-seconds, examination of the ranges that we have developed in Table 5.5 shows that there is just one possible time for which (3) is satisfied (subject to H→M→S):

9 hr., 5 min., $25\frac{1}{2}$ sec.

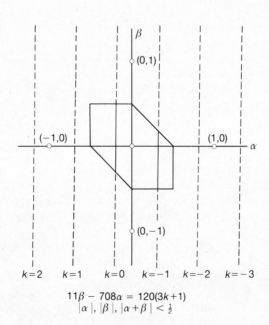

$$11\beta - 708\alpha = 120(3k+1)$$
$$|\alpha|, |\beta|, |\alpha+\beta| < \tfrac{1}{2}$$

Diagram 5.3

Table 5.5

		Limits of α		$\|\alpha\|, \|\beta\|, \|\alpha+\beta\| \leq 1$		
k	m	from	to	at time	from	to
-1	5	$\frac{469}{1416}$	$\frac{491}{1438}$	5 hr., 49 min., $+$	$9\frac{4}{59}$ sec.	$9\frac{129}{719}$ sec.
0	8	$-\frac{251}{1438}$	$-\frac{229}{1416}$	9 hr., 5 min., $+$	$25\frac{265}{719}$ sec.	$25\frac{30}{59}$ sec.

We now need to allow for the possibility that the clockwise cyclic order of the hands round the clock is H→S→M: doing so adds to our list the 'reflection' time:

$$\text{2 hr., 54 min., } 34\tfrac{1}{2} \text{ sec.}$$

Since we know that Dick's watch stops between noon and 6.00 p.m., it follows that this is the unique answer to the problem.

Particular solution III

When Harry's watch stops (between noon and 6.00 p.m.), the angles the three hands make with each other all differ from 120° by no more than iota degrees—and 'iota' is the smallest number for which the situation is possible at all.

Following the lines of the General Solution, our first question is: 'What is the smallest value of δ that allows (7) to have a solution?' If $\delta < \frac{120}{719}$, (7) has no solution; but if $\delta = \frac{120}{719}$, (7) does have a solution: $k = 0$. So:

$$\text{iota} = \tfrac{120}{719}.$$

The relevant part of Table 5.1 is Table 5.6.

We now sketch (3) (with $\delta = \frac{120}{719}$) and (6) in Diagram 5.4.[10] From (3), (6) (keeping an eye on Diagram 5.4), and then (1), we have Table 5.7.

We now need to allow for the possibility that the clockwise cyclic

---

[10] There is no real need to do so, but . . . In the two previous problems (with $\delta = 1$ and $\delta = \frac{1}{2}$), the limits of possible α for $k=0$ were given by $\alpha + \beta = -\delta$ and $\beta = \delta$ (in conjunction with (6)). There is a possible risk that we might fall into the error of assuming that this will continue to apply as δ continues to decrease. (It does continue to apply—until $\delta = \frac{120}{708}$; thereafter, as δ continues to decrease, the limits are given by $\alpha = -\delta$ and $\beta = \delta$.)

Table 5.6

k	m	n
0	8	536

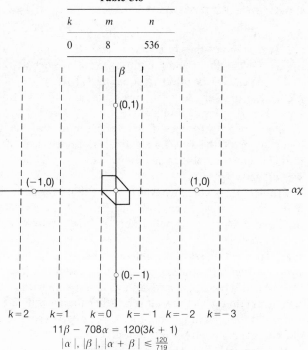

$$11\beta - 708\alpha = 120(3k + 1)$$
$$|\alpha|, |\beta|, |\alpha + \beta| \leq \tfrac{120}{719}$$

Diagram 5.4

Table 5.7

k	m	α	$|\alpha|, |\beta|, |\alpha+\beta| \leq 1$ at time
0	8	$-\tfrac{120}{719}$	9 hr., 5 min., $25\tfrac{325}{719}$ sec.

order of the hands round the clock is H→S→M: doing so adds to our list the 'reflection' time:

$$\text{2 hr., 54 min., } 34\tfrac{394}{719} \text{ sec.}$$

Since we know that Harry's watch stops between noon and 6.00 p.m., it follows that this is the time at which it stopped.[11]

[11] If we calculate the positions of the hands at this time, we get

$$\text{H @ } 87\tfrac{207°}{719}, \quad \text{M @ } 327\tfrac{327°}{719}, \quad \text{S @ } 207\tfrac{207°}{719};$$

and so

$$\alpha = -\tfrac{120}{719}, \qquad \beta = \tfrac{120}{719}, \qquad \gamma = 0.$$

Composer's problem

A classic Clock Problem (with a multitude of variations) is: 'When are the two hands—the hour-hand and the minute-hand—exactly opposite each other?' We can rephrase that: 'When do the two hands divide the clock-face into two equal parts?' (As you well know, there are eleven answers in every twelve hours: any angular situation repeats itself every $\frac{12}{11}$ hours.)

'Aha,' one says to oneself, 'let's bring in a third (i.e. the second-) hand. When do the *three* hands divide the clock-face into *three* equal parts?' Unfortunately the answer is: 'Never—well, never exactly. But we can get very close.'

Problem III—which was the first one that I thought about—looks at the question of how close one can get. In its original form I agonized over different definitions of 'how close?'[12] But eventually sanity prevailed, and I realized that nobody would want a problem that depended on (say) the difference between $\frac{285998}{509173}$ and $\frac{802}{1427}$. So I firmly settled on minimizing the maximum of $|\alpha|, |\beta|, |\gamma|$.

Problem III has the disadvantage (compared with two-hand clock problems) that it's a 'how close?' rather than a 'where exactly?' one; but it has the advantage that there are only two solutions, instead of eleven, in each twelve-hour period.

Problem III has another disadvantage—its solution involves rather nasty fractions (any fraction with a denominator 719 is a rather nasty one); and watches do not normally stop after multiples of $\frac{1}{719}$ second.[13] That disadvantage is removed very simply, by asking oneself (or a horologist) the question: 'When *do* they stop?' The answer to that question leads immediately to Problems I and II.

Half-second jerks (Problem II) lead—as one would expect—to an answer very close to the Problem III solution. It was with some satisfaction that I realized that whole-second jerks (Problem I) led to

---

[12] There are quite a number of possible 'how close' criteria, among them:

$\min(\alpha^2 + \beta^2 + \gamma^2)$	occurs at 2 hr., 54 min., $34\frac{285998}{509173}$ sec.
$\left.\begin{array}{l}\min(|\alpha|+|\beta|+|\gamma|) \\ \min \max(|\alpha|, |\beta|, |\gamma|)\end{array}\right\}$	occurs at 2 hr., 54 min., $34\frac{394}{719}$ sec.
$\min \max(\alpha, \beta, \gamma)$	occurs at 2 hr., 54 min., $34\frac{802}{1427}$ sec.
$\max \min(\alpha, \beta, \gamma)$	occurs at 2 hr., 54 min., $34\frac{38}{73}$ sec.

(and at their reflections).

[13] Nor do they normally stop after multiples of $\frac{1}{11}$ second: see Appendix.

an answer far removed from either. (That Problems I and II lead to such disparate answers is my reason—and, I hope, justification—for including both of them.)

But none of this addresses my justification for having included these three problems at all. They are extremely 'elementary'—in that they are solvable using processes familiar to any numerate primary schoolchild. But they are not particularly 'easy': one needs to stay in control of the arithmetic, and to think ahead about the number and kind of particular cases to be considered. It is for that reason that I have not only included them but have also gone into considerable (possibly excessive) detail in the General Solution.[14]

Appendix: Two-handed problems

When the second-hand of Tom's three-handed watch jerks forward, the jerk is pretty obvious—after all, the hand moves through $6°$. What are not so obvious are the accompanying jerks of the minute-hand (through $\frac{1}{10}°$) and of the hour-hand (through $\frac{1}{120}°$). But these jerks do take place: and they take place whether or not the watch, or clock, has a second hand at all.

All clocks and watches 'beat': the hands advance in jerks.[15] Some beat in seconds, some in half-seconds, some in much smaller fractions of a second.[16] But all beat—and none, as far as I can discover, has yet been made that beats in eleventh-seconds.

One consequence is that a problem such as

'My clock stopped somewhere between 7 and 8 o'clock with its hands diametrically opposite each other. At what time did it stop?'

lays itself open to the counter-question

'At what rate does your clock beat? Sixty times a minute? Three hundred times a minute? Or what?'

If the answer to the counter-question is either

'$11j$ times a minute' (for any integer j)

---

[14] I am still not satisfied with the line of attack that I have adopted in the General Solution: there should be a better way!

[15] So Harry's 'no jerk' watch in Problem III is, unfortunately, a figment of my imagination: no such watch exists.

[16] And some—old long-pendulum clocks—beat every $\frac{5}{4}$ seconds. (A fact that is totally irrelevant to this discussion!)

or

'It doesn't beat: the hands move smoothly, with no jerks',

then the answer to the original problem is

'At $5\frac{5}{11}$ minutes past 7' (with, perhaps, the addition 'and your clock is almost certainly unique').

But if the answer to the counter-question is any number that does not have 11 as a factor, then the 'correct' answer to the original problem is

'It can't happen.'

So the problem-setter needs to assume a hypothetical clock that beats $11k$ times a minute (or one that doesn't beat at all).

That introduces another—very minor—difficulty. With such a clock the relative angular position of the hands is repeated every $\frac{12}{11}$ hours, so that the problem has, basically, eleven possible solutions: an additional constraint (e.g. 'sometime between 7 and 8 o'clock') has to be introduced to render the solution unique.

These two reservations about the typical classical two-handed problems may be—and perhaps should be—dismissed as pedantic quibbles. But the search to circumvent them has its uses. It leads us to a different *kind* of problem: one that uses the beat phenomenon as an advantage (and, incidentally, reduces the eleven-fold solution to a two-fold or, at worst, three-fold one).

EXAMPLE: 'My clock (which beats in seconds) stopped at some time between noon and 5.00 p.m.[17] with its hands as near to being diametrically opposite to each other as they possibly could be. What time did it stop at?'

ANSWER: 'At $38\frac{11}{60}$ minutes after 1 o'clock (when the angle between the hands differs from $180°$ by $\frac{1}{120}°$).'

In this kind of problem, not only can we assume realistic clocks (ones that beat with frequencies that occur in practice), but we also have the advantage—from a problem-setter's point of view—that there is an additional parameter (the beat frequency) at our disposal.

---

[17] We need to exclude 6.00 p.m. (when the hands are exactly diametrically opposite); and we need to specify one or other of the two periods 0.00–6.00 and 6.00–12.00 (since any solution in one of these periods has a corresponding, mirror-image, solution in the other.)

6

MAGIC BONDS

Problem

Each of James's nine children had a prime number of Premium Bonds. All nine together had less than 24,579.[1] Brian and Caroline together had as many as Georgina and Harry together; Eleanor and Imogen together had as many as Brian and Francis together; and Brian and Imogen together had as many as David and Eleanor together. The children joined in buying a Christmas present for their father: to raise the money, each cashed a prime number of his or her Premium Bonds and kept the rest. Each kept a prime number of Premium Bonds. No two kept the same number, but David and Harry together kept as many as Caroline and Georgina together; Eleanor and Francis together kept as many as Harry and Imogen together; and Eleanor and Georgina together kept as many as Anne and Brian together.

How many did Anne keep?

[1] A *lot* less, actually. (Why did I choose 24,579? Because if I'd chosen any larger number, the problem wouldn't have a unique solution.)

Solution

1   There cannot be two (or more) children who kept 2 bonds, since no two kept the same number. There cannot be just one child who kept 2 bonds, because each of the other eight would then have kept an odd (odd, because prime-other-than-2) number of bonds, and as all nine of the children are involved in the equalities that occur after the bond-cashing, at least one of those equalities would be impossible. So each of the nine children kept an odd prime number of bonds.

But each child started with a prime number of bonds. And 2 is the only prime that, deducted from a prime, can leave an odd prime. So each child cashed 2 bonds.

2   Consequently the equalities that occurred before the bond-cashing also occurred after it. So we have six equalities:

$$B + C = G + H \qquad E + I = B + F$$
$$B + I = D + E \qquad D + H = C + G$$
$$E + F = H + I \qquad E + G = A + B$$

where A is the number of bonds that Anne kept (and so on).

It is the value of A that we want; so we express A in terms of the other quantities. We get

$$2A = B + F = C + G = D + H = E + I.$$

3   The number of bonds that each child kept is not only a prime but is also 2 less than a prime: so the nine different numbers of bonds kept must be among

$$3, 5, 11, 17, 29, 41, 59, 71, 101, 107, 137, 149,$$
$$179, 191, 197, 227, 239, 269, 281, 311, 347, \ldots$$

A must be not only a member of this set but also (by para. 2) the average of two other members of the set in (at least) four different ways.

By trial (not too arduous) we find that 149 is the least number that meets that requirement.

$$(2 \times 149 = 17 + 281 = 29 + 269 = 59 + 239 =$$
$$71 + 227 = 101 + 197 = 107 + 191.)$$

Trial (again not too arduous) then shows that the requirements are all met by (for example)

$$
\left.\begin{array}{ccc}
\text{I} & \text{B} & \text{C} \\
\text{H} & \text{A} & \text{D} \\
\text{G} & \text{F} & \text{E}
\end{array}\right\} \equiv \left\{\begin{array}{ccc}
59 & 197 & 191 \\
281 & 149 & 17 \\
107 & 101 & 239
\end{array}\right.
$$

where the layout of the solution reflects the fact that the holdings of bonds (whether before or after the bond-cashing) form a Magic Square (as suggested by the title of the problem).

So we have a solution: A = 149. (And we also have that all nine children together started with $9 \times 151 = 1{,}359$ bonds, which, as promised, is a lot less than 24,579.)

4   We also know that it is the smallest solution. But we do not know that it is the *only* solution (the only solution, that is, in which all nine children together start with less than 24,579 bonds). Proving this is distinctly more arduous than what we have done so far.

As a first step, it is convenient to prove (it needs proof: we shouldn't just assume it) that the six equalities listed at the start of para. 2 do in fact imply that the nine quantities satisfying them must form a Magic Square, with A as the central element. This is straightforward.

We then examine in turn (in ascending order) the members of the set given in para 3, first to see which of them can be expressed as the average of two other members of the set in (at least) four different ways. Those (after 149) that can be so expressed are

$$
2 \times 269 \quad = 17 + 251 \ = 107 + 431 = 191 + 347 = 227 + 311,
$$

$$
2 \times 419 \quad = 29 + 809 \ = 179 + 659 = 197 + 641 = 239 + 599 = 269 + 569,
$$

$$
\vdots \qquad\quad \big\} \ \ \text{thirty-two others}
$$

$$
2 \times 2729 \quad = 17 + 5441 = 41 + 5417 = 179 + 5279 =
$$

$$
227 + 5231 = 659 + 4799 = 809 + 4649
$$

$$
= 821 + 4637 = 1229 + 4229 = 1301 + 4157 =
$$

$$
1607 + 3851 = 1787 + 3671
$$

$$
= 1877 + 3581 = 1931 + 3527 =
$$

$$
1997 + 3461 = 2087 + 3371
$$

$$
= 2339 + 3119 = 2657 + 2801.
$$

. . .

The next step is to see if, from the numbers so obtained, we can construct a

Magic Square. (This is not so fearsome a task as may appear at first sight: there are several properties of Magic Squares that can be used to reduce the labour very considerably. For example, the next-to-smallest element of a Magic Square is always the average of the smallest element and another element.)

That step takes some time. But eventually we find that there is no case in which we can construct a Magic Square—until we come to consider the 2729 possibility. And there we find that

$$
\begin{array}{ccc}
659 & 4157 & 3371 \\
5441 & 2729 & 17 \\
2087 & 1301 & 4799
\end{array}
$$

meets the requirements. But these figures, representing the numbers of bonds that the children kept, imply that together they started with $9 \times 2731 = 24{,}579$ bonds. And that, according to the conditions of the problem, is (just) too many.

Composer's problem

1 In 1973, looking for a possible problem, I was constructing Magic Squares of order 3 having each element prime. The drawback (from a problem-composition point of view) was that there were too many of them: in order of increasing central element, I had

$$
\begin{array}{ccc}
17 & 89 & 71 \\
113 & 59 & 5, \\
47 & 29 & 101
\end{array}
\qquad
\begin{array}{ccc}
41 & 89 & 83 \\
113 & 71 & 29, \\
59 & 53 & 101
\end{array}
\qquad
\begin{array}{ccc}
37 & 79 & 103 \\
139 & 73 & 7, \ldots; \\
43 & 67 & 109
\end{array}
$$

or (putting it more briefly—listing just the central elements)

59, 71, 73, 89, 103, 109, 127 (two of these), 131, 137, 139, 149, 151, 157, 167 (two of these), 173, . . .

My next thought was to try to compose a problem round the 'two of these' situations: but there were still rather too many of them (with central elements 127, 167, 227, . . .).

Then I noticed that the '149' and '151' magic squares differed by 2 not just in their central elements, but in every one of their elements:

$$
\begin{array}{ccc}
59 & 197 & 191 \\
281 & 149 & 17, \\
107 & 101 & 239
\end{array}
\qquad
\begin{array}{ccc}
61 & 199 & 193 \\
283 & 151 & 19. \\
109 & 103 & 241
\end{array}
$$

This looked more promising. But there was still a question to be answered: when did it happen again? Finding out took some time,[2] but eventually I had that the next time it happened was

659	4157	3371		661	4159	3373
5441	2729	17,		5443	2731	19.
2087	1301	4799		2089	1303	4801

The occurrence seemed sufficiently rare to be interesting.

2   The next thing to do was to find a context for the problem: preferably one in which the context itself would unobtrusively but inevitably confine the number of possible solutions to one.

The context of Premium Bond holdings seemed ideal; in the mid-seventies the maximum number of Premium Bonds that an individual could legally hold was 2,000. (The limit was raised in September 1979 to 3,000.) So I wrote a problem-statement that started:

> *Each of James's nine law-abiding children had a prime number of Premium Bonds. Brian and Caroline together . . .*

(with no need to specify any upper limit explicitly: 'law-abiding' was enough in itself).

It was *very* shortly after publication[3] that it was pointed out to me that the problem, as set, did *not* have a unique solution. What I had missed was that the limit on the number of Premium Bonds that an individual could legally hold had been raised, in April 1980, to 10,000.

So, for publication here, I have had to introduce an explicit upper limit in order to restore uniqueness.

(I *do* wish that the authorities had left the legal limit at 2,000 or 3,000: I was rather fond of 'law-abiding' as a mathematical restraint!)

---

[2] I sketch the method in para. 4 of the Solution section.
[3] First published (in the 'law-abiding' form) on 21 December 1986 as Brainteaser 1268 in *The Sunday Times Magazine*.

7

TWICE THE SIZE

Problem I[1]

'My lawn is triangular; each side is an exact number of yards long; the number of square yards in its area is exactly twice the number of yards in its perimeter.'

That's what Tom said; it's what Dick said; it's what Harry said.

(If you really want to know: Tom's lawn's sides are 13, 14, 15 yards long; so its area, in square yards, is

$$\tfrac{1}{4}\sqrt{\{(13+14+15)(-13+14+15)(13-14+15)(13+14-15)\}},$$

which is 84, which is twice (13+14+15). But Tom's lawn is just an example: you can forget about this bit in brackets from now on.)

The interesting thing is that the areas of Dick's lawn and Harry's lawn are equal—but the longest side of Dick's lawn is longer than the longest side of Harry's lawn.

So what are the lengths, in yards, of the sides of Harry's lawn?

[1] First published on 29 March 1987 as Brainteaser 1282 in *The Sunday Times Magazine*.

Problem II

'*My* lawn', said Anne, 'is triangular; each side is an exact number of yards long; and the number of square yards in its area is a rational fraction, less than unity, of the number of yards in its perimeter. Even if I told you what the fraction was, you wouldn't be able to work out the area of my lawn—there'd be two possible answers, one twice the size of the other. So what is the fraction?'

Initial discussion

1   A triangle has sides a, b, c yards long, and area Δ square yards, where a, b, c, Δ are integers;

$$4\Delta = \sqrt{\{(a+b+c)(-a+b+c)(a-b+c)(a+b-c)\}}. \tag{1}$$

Without loss of generality, we can assume that

$$a \geqq b \geqq c. \tag{2}$$

If $a+b+c$ were odd then the right-hand side of (1) would be odd. But the left-hand side of (1) is even. So there exist integers p, q, r such that

$$-a+b+c = 2p, \tag{3a}$$

$$a-b+c = 2q, \tag{3b}$$

$$a+b-c = 2r. \tag{3c}$$

(1) becomes

$$\Delta = \sqrt{\{(p+q+r)pqr\}}; \tag{4}$$

and the perimeter of the triangle is P yards, where

$$P = 2(p+q+r). \tag{5}$$

From (2), (3) we have

$$p \leqq q \leqq r. \tag{6}$$

2   Suppose that we are told that the ratio of Δ to P is f/g, where f, g are integers. We can assume that f/g is in its lowest terms: that is, that

$$\mathrm{hcf}(f, g) = 1. \tag{7}$$

By (4), (5) we then have

$$g^2 pqr = 4f^2(p+q+r). \tag{8}$$

3   There are (at least) two ways in which we can proceed from (8): one of them appropriate to the situation in which we are told the *specific* numeric values of f, g; the other more appropriate to the situation in which we are told only that f/g is less than some specified quantity. The first is illustrated in the (first part of) Particular Solution I; the second is given in the (first part of) Particular Solution II.

Particular solution I

1   In Problem I we are given (in the terminology of the Initial Discussion) that $f/g = 2$; so, from (8), we have

$$pqr = 16(p + q + r). \tag{9}$$

It follows from (6) that $p \leq 6$. We now consider in turn the values 6, 5, 4, 3, 2, 1 of p.

1.1   *p = 6*

When $p = 6$, (9) becomes $(3q - 8)(3r - 8) = 2^4 \times 13$, of which there is just one solution satisfying (6):

$3q - 8$	$3r - 8$	q	r	Δ
13	16	7	8	84

1.2   *p = 5*

When $p = 5$, (9) becomes $(5q - 16)(5r - 16) = 2^4 \times 41$, of which there are no solutions satisfying (6).

1.3   *p = 4*

When $p = 4$, (9) becomes $(q - 4)(r - 4) = 2^5$, of which there are three solutions satisfying (6):

$q - 4$	$r - 4$	q	r	Δ
4	8	8	12	96
2	16	6	20	120
1	32	5	36	180

1.4   *p = 3*

When $p = 3$, (9) becomes $(3q - 16)(3r - 16) = 2^4 \times 5^2$, of which there are four solutions satisfying (6):

$3q - 16$	$3r - 16$	q	r	Δ
20	20	12	12	108
8	50	8	22	132
5	80	7	32	168
2	200	6	72	324

1.5   *p = 2*

When $p = 2$, (9) becomes $(q - 8)(r - 8) = 2^4 \times 5$, of which there are five solutions satisfying (6):

$q-8$	$r-8$	q	r	Δ
8	10	16	18	144
5	16	13	24	156
4	20	12	28	168
2	40	10	48	240
1	80	9	88	396

1.6 *p = 1*

When $p=1$, (9) becomes $(q-16)(r-16) = 2^4 \times 17$, of which there are five solutions satisfying (6):

$q-16$	$r-16$	q	r	Δ
16	17	32	33	264
8	34	24	50	300
4	68	20	84	420
2	136	18	152	684
1	272	17	288	1224

2   There are thus eighteen distinct integer solutions of (9) satisfying (6).

Among them there are just two that have the same value of Δ:

Δ		p	q	r		a	b	c
168	$\begin{cases} \\ \end{cases}$	3	7	32		39	35	10
		2	12	28		40	30	14

(where the values of a, b, c are given by (3)). So Dick's lawn has sides 40, 30, 14 yards, and Harry's lawn has sides 39, 35, 10 yards.[2]

Particular solution II

1   In Problem II we are given (in the terminology of the Initial Discussion) that

$$f/g < 1. \tag{10}$$

It follows from (6), (8) that

$$p \leqq 3. \tag{11}$$

We consider in turn the values 3, 2, 1 of p.

---

[2] Each of their lawns has area 168 square yards—twice the size of Tom's lawn. But the only relevance of that is that it (partly) accounts for the title of the problem.

2   p = 3

When $p = 3$ (8) becomes

$$3g^2qr = 4f^2(q + r + 3). \tag{12}$$

It follows from (6), (10) that $q \leq 3$, and, so, that $q = 3$. Hence (12) becomes

$$9g^2r = 4f^2(r + 6),$$

which has no integer solutions satisfying (6).

3   p = 2

When $p = 2$ (8) becomes

$$g^2qr = 2f^2(q + r + 2). \tag{13}$$

It follows from (6), (10) that $q \leq 4$, and, so, that $q = 4, 3,$ or 2. We now consider in turn these values of q.

3.1   When $q = 4$ (13) becomes

$$2g^2r = f^2(r + 6),$$

which has no integer solutions satisfying (10).

3.2   When $q = 3$ (13) bcomes

$$3g^2r = 2f^2(r + 5),$$

which has just one integer solution satisfying (6), (7), (10): namely, $r = 3$, $f = 3$, $g = 4$.

3.3   When $q = 2$ (13) becomes

$$g^2r = f^2(r + 4),$$

which has no integer solutions.

Hence (13) has just one integer solution satisfying (6), (7), (10):

$$p, q, r = 2, 3, 3; \qquad f/g = 3/4.$$

4   p = 1

When $p = 1$ (8) becomes

$$g^2qr = 4f^2(q + r + 1). \tag{14}$$

It follows from (6), (10) that $g \leq 8$. We now consider in turn these values of q.

4.1   When $q = 8, 7, 6, 5$ we find (by arguments similar to those of Section 3) that there are no solutions satisfying (6), (10).

4.2   When $q = 4$ (14) becomes

$$g^2 r = f^2(r+5),$$

which has just one integer solution satisfying (6), (7), (10): namely, $r = 4$, $f = 2$, $g = 3$.

4.3   When $q = 3$ (14) becomes

$$3g^2 r = 4f^2(r+4). \tag{15}$$

[Here we have a new difficulty. In all the cases that we have considered so far there have been finite numbers of values of r to consider (limited by (10))—or, at worst, we have had the special cases considered in paras. 3.3 and 4.2. But (10) provides no limit when we come to (15): at first sight it seems that we shall have to consider *all* values of r (≥ 3).]

4.31   We first of all observe from (15) that r must be a factor of $16f^2$, and that (by (7)) f^2 must be a factor of $3r$. So there exist integers m, n such that $16f^2 = mr$ and $3r = nf^2$. It follows that $mn = 48$. There are ten possible (ordered) pairs m, n that have the product 48: we need to consider each of them. (Luckily, seven of them can be disposed of almost immediately—we shall do that in para. 4.32: they are distinguished by an asterisk (*) in the following table.)

Table 7.1

m	n	from (15)	
48	1	$3g^2 = 4f^2 + 48$	*
24	2	$3g^2 = 4f^2 + 24$	*
16	3	$3g^2 = 4f^2 + 16$	*
12	4	$3g^2 = 4f^2 + 12$	
8	6	$3g^2 = 4f^2 + 8$	
6	8	$3g^2 = 4f^2 + 6$	*
4	12	$3g^2 = 4f^2 + 4$	*
3	16	$3g^2 = 4f^2 + 3$	
2	24	$3g^2 = 4f^2 + 2$	*
1	48	$3g^2 = 4f^2 + 1$	*

4.32   Consider the equation in the first entry in Table 7.1:

$$3g^2 = 4f^2 + 48.$$

Clearly, g must be even; so we can write $g = 2G$, and have

$$3G^2 = f^2 + 12. \tag{16}$$

By (7) we have that f must be odd: and so the right-hand-side of (16) must leave remainder 5 on division by 8. But the left-hand-side of (16) can leave only remainder 0, 3, or 4 on division by 8. So (16) has no integer solution.

Similar considerations show the impossibility of the other asterisked equations in the table.

4.33   When $m = 12$ and $n = 4$, we have

$$3g^2 = 4f^2 + 12, \tag{17a}$$

$$3r = 4f^2. \tag{17b}$$

(17a) has an infinity of solutions, but luckily we do not need to develop them here.[3] For the purpose of answering Anne's question, we need to note only those with $f = 2$ or $f = 3$:[4] there is just one:

$$f = 3, \qquad g = 4.$$

4.34   When $m = 8$ and $n = 6$, we have

$$3g^2 = 4f^2 + 8, \tag{18a}$$

$$r = 2f^2. \tag{18b}$$

(18a) has an infinity of solutions: but none has $f = 2$ or $f = 3$.

4.35   When $m = 3$ and $n = 16$, we have

$$3g^2 = 4f^2 + 3, \tag{19a}$$

$$3r = 16f^2. \tag{19b}$$

(19a) has an infinity of solutions, but none has $f = 2$ or $f = 3$.

It follows that (15) has just one integer solution having $f = 2$ or $f = 3$: namely, $r = 12, f = 3, g = 4$ (a result that we could have obtained rather more shortly); and that all the other integer solutions of (15) that satisfy (6), (7), (10) also satisfy one or other of (17a), (18a), (19a) (a result that we shall need later, and which was the reason for the rather lengthy detail into which we have entered).

[3] (17a) (when we have written $f = 3F$, $g = 2G$, as clearly we can, to get $G^2 = 3F^2 + 1$) is an example of Pell's Equation ($x^2 - Dy^2 = 1$). I am resisting the temptation to discuss Pell's Equation here: we don't need that discussion for this particular problem. But you might like to follow it up.

[4] I shall—I hope—justify this assertion in Section 5!

4.4   When $q = 2$, (14) becomes

$$g^2r = 2f^2(r+3). \tag{20}$$

[We have the same difficulty to contend with as in para. 4.3.]

4.41   We first of all observe from (20) that r must be a factor of $6f^2$, and that (by (7)) f^2 must be a factor of r. So there exist integers m, n such that $6f^2 = mr$ and $r = nf^2$. It follows that $mn = 6$. There are four possible (ordered) pairs m, n that have the product 6: we need to consider all of them (see Table 7.2).

Table 7.2

m	n	from (20)	
6	1	$g^2 = 2f^2 + 6$	*
3	2	$g^2 = 2f^2 + 3$	*
2	3	$g^2 = 2f^2 + 2$	
1	6	$g^2 = 2f^2 + 1$	

4.42   Consider the equation in the first entry in Table 7.2:

$$g^2 = 2f^2 + 6.$$

In it (by (7)) neither f nor g is divisible by 3; so the left-hand-side must leave remainder 1 on division by 3, and the right-hand-side remainder 2, which is impossible. So the equation has no integer solution.

Similar considerations show the impossibility of the other asterisked equation in the table.[5]

4.43   When $m = 2$ and $n = 3$, we have

$$g^2 = 2f^2 + 2, \tag{21a}$$

$$r = 3f^2. \tag{21b}$$

(21a) has an infinity of solutions: but none has $f = 2$ or $f = 3$.

4.44   When $m = 1$ and $n = 6$, we have

$$g^2 = 2f^2 + 1, \tag{22a}$$

$$r = 6f^2. \tag{22b}$$

[5] We could equally well have shown the impossibility of the two asterisked equations by considering remainders on division by 8, as in para. 4.32.

(22a) has an infinity of solutions: there is just one with $f=2$ or $f=3$:

$$f=2, \qquad g=3.$$

It follows that (20) has just one integer solution having $f=2$ or $f=3$, namely

$$r=24, \qquad f=2, \qquad g=3,$$

and that all the other integer solutions of (20) that satisfy (6), (7), (10) also satisfy one or other of (21a), (22a).

4.5   When $q=1$ (14) becomes

$$g^2r=4f^2(r+2),$$

which has no integer solutions.

Hence (14) has three integer solutions having $f=2$ or $f=3$ and satisfying (6), (7), (10):

$$p, q, r = 1, 4, 4, \qquad f/g = 2/3;$$
$$p, q, r = 1, 3, 12 \qquad f/g = 3/4;$$
$$p, q, r = 1, 2, 24, \qquad f/g = 2/3;$$

and all other solutions satisfying (6), (7), (10) satisfy one or other of (17a), (18a), 19a), (21a), (22a) and do not have $f=2$ or $f=3$.

5   We now collect the results of Sections 1–4. We have that the solutions of (8) that satisfy (6), (7), (10) are

(α)	p	q	r	f	g
	2	3	3 ⎫		
	1	3	12 ⎭	3	4

(β)	p	q	r	f	g
	1	4	4 ⎫		
	1	2	24 ⎭	2	3

(γ)   and an infinity of others; each satisfying one of (17a), (18a), 19a), (21a), (22a); none having $f=2$ or $f=3$.

Is there any other (other, that is, than (α) or (β)) pair of distinct solutions that share the same value of f/g? If there is, the two solutions must both be members of (γ). But on taking in turn each possible pair of the equations (17a), (18a), (19a), (21a), (22a), we find that the only two that can share the same (rational, non-zero) value of f/g are (18a) and (21a), with $f/g = 1/2$; and in that case (18) gives $q = 3$, $r = 2$, contrary to (6). So (α) and (β) are the only pairs of distinct solutions that share the same value of f/g.

The two solutions (β) share the same value of f/g; but the associated areas are (using (4)) 12 and 36 square yards: *not* 'one twice the size of the other'.

The two solutions (α) share the same value of f/g; and the associated areas are (using (4)) 12 and 24 square yards: one twice the size of the other.

So the number of square yards in the area of Anne's lawn is 3/4 of the number of yards in its perimeter.[6]

Composer's problem

When 'the number of square yards in the area equals the number of yards in the perimeter', we have (in the terminology of the Initial Discussion)

$$pqr = 4(p + q + r).$$

This has five distinct (non-zero, integer) solutions:

Δ	p	q	r	a	b	c
24	2	4	6	10	8	6
30	2	3	10	13	12	5
36	1	8	9	17	10	9
42	1	6	14	20	15	7
60	1	5	24	29	25	6

I used this as the basis for a problem in an earlier book;[7] but at the time I was thinking too much about finding an interesting context for it, and not enough about possible generalizations. Later, however, I realized that there was a wealth of problem material in

$$g^2 pqr = 4f^2(p + q + r)$$

(equation (8) of the Initial Discussion), ranging from

---

[6] Anne's lawn either has area 12 square yards and sides 6, 5, 5 yards, or has area 24 square yards and sides 15, 13, 4 yards.

[7] 'Counting Sheep', in *Mathematical Byways in Ayling, Beeling, and Ceiling* (Oxford University Press, 1984).

'What is the least possible value of f/g for which there is any (non-zero, integer) solution at all?'[8]

through the two problems that I have set here, and

'What is the nearest that f/g can get to unity, without actually being unity?'[9]

to ones involving the solution of Pell's Equation,

$$x^2 - Dy^2 = 1,$$

and ones involving the solution of Pell-like equations such as

$$x^2 - 3y^2 = -2$$

(*vide* para. 4.34 of Particular Solution II), but more difficult than that particular example.

My problem was, mainly, one of selection from among these various possibilities. I did not want to include more than two examples; I wanted those examples that I did include to illustrate sensibly different approaches; I wanted a degree of difficulty that was (roughly) consistent with other problems in this book; and I wanted to be able to set the problems in contexts that were clear and fairly concise.

In an earlier draft I set out to include

'What is the nearest that f/g can get to unity, without actually being unity?'

but failed to find a clear concise context for it.

(The two problems on which I eventually settled at least have the trivial merit of being able to share the same title!)

---

[8] Answer: 1/2.
[9] Answer: 6/7 below; 10/9 above.

8

ALPHAMETICS

Each of the problems in this chapter is independent of the others.
  In each problem:
    Capital letters stand for digits; different capital letters stand for
    different digits; the same capital letter stands for the same digit
    wherever it occurs; and no number starts with a zero.

Problem I: STAND STILL, DAMNIT[1]

If STAND + STILL = DAMNIT, what is the RESTRAINT?

Problem II: UNDERWATER ERROR[2]

If UNDER + WATER + ERROR = DROWN, what is UNINTENDED?

Problem III: ALICE[3]

ALICE is odd; she turned ROUND and took the result away from
    herself to get ROUND again. So what is the CLUE?

Problem IV: REINDROP[4]

If RIDER + DROPS + REINS = LOSES − HORSE, what is the HORSE?

---

[1] First published (edited) as Braintwister 486 in Douglas Barnard's column in the
Daily Telegraph.
[2] First published (considerably edited) as Braintwister 421 in Douglas Barnard's
column in the *Daiy Telegraph*.
[3] First published (very considerably edited!) as Braintwister 440 in Douglas
Barnard's column in the *Daily Telegraph*.
[4] First published in the December 1979 issue of the *Hawley Riding Club Newsletter*.

Problem V: THF[5]

If TRUST + HOUSE + FORTE = HOTELS, what is RESTFULNESS?

[5] Not previously published: Trust House Forte didn't want to know!

Particular solution I

We are told

$$
\begin{array}{r}
\text{S T A N D} \\
+ \quad \text{S T I L L} \\
\hline
= \text{D A M N I T}
\end{array}
$$

Clearly,

$$D = 1. \tag{1}$$

We then have

$$1 + L = T + 10\alpha, \tag{2}$$

$$N + L + \alpha = I + 10\beta, \tag{3}$$

$$A + I + \beta = N + 10\gamma, \tag{4}$$

$$2T + \gamma = M + 10\delta, \tag{5}$$

$$2S + \delta = A + 10; \tag{6}$$

where α, β, γ, δ, (the 'carries') are each either 0 or 1.

If $\alpha = 1$ we would have from (2) that $L = 9$, and so from (3) that $N = I$ or $N = I - 10$, neither of which is possible. So

$$\alpha = 0. \tag{7}$$

Adding (3) and (4) (and using (7)), we have

$$A + L = 9\beta + 10\gamma, \tag{8}$$

from which it follows that β, γ cannot both be 0, and cannot both be 1. That leaves two possibilities:

$$
\text{or} \quad
\left.
\begin{array}{ll}
\beta = 1, & \gamma = 0; \\
\beta = 0, & \gamma = 1.
\end{array}
\right\}
\tag{9}
$$

(We could at this stage proceed by enumeration of cases: for example, we could let A take in turn the values $0, 1, 2, \ldots, 9$; for each then calculate S and δ from (6); then L (two possible values, because of (9)) from (8); then T from

(2); and so on. We would have a table of twenty lines, from which we would then discard those that involved two different letters having the same numerical value. This is perfectly feasible; but I prefer to defer 'enumeration of cases' as long as possible, and in this particular problem there is a further step we can take before being forced into it.)

By (2), (6), (7) we have

$$A + L = 2S + T + \delta - 11,$$

and so, by (5), (8),

$$4S + M = 18\beta + 21\gamma - 12\delta + 22. \tag{10}$$

((10) is useful because of the coefficient 4 of S; corresponding to a given value of the right-hand-side there can be at most three possible pairs of values of S, M. Our table need now, in fact, occupy only eight lines.)

We develop Table 8.1 by taking the four possible combinations of β, γ, δ; then, using (10), we deduce the possible values of S and M: there are, in all, eight possibilities. For each we then deduce in turn T (by (5)), L (by (2)), and A (by (6)). In each line we stop if and when we arrive at a duplication (*).

Table 8.1

D	α	β	γ	δ	S	M	T	L	A
					5	8	9	8*	
		1	0	1	6	4	7	6*	
					7	0	5	4	5*
1	0	0	1	1	6	7	8	7*	
					7	3	6	5	5*
		1	0	0	8	8*			
					9	4	2	1*	
		0	1	0	9	7	3	2	8

From this we have that there is just one possible set of values of α, β, γ, δ; D, S, M, T, L, A   (0, 0, 1, 0; 1, 9, 7, 3, 2, 8).

This means that just the digits 0, 4, 5, 6 are available for N and I. But only two of them differ by 2; and from (3) we have that

$$I - N = 2.$$

Consequently

$$I = 6, \qquad N = 4.$$

There are still two letters left whose values have to be determined: R and E (which occur in RESTRAINT). There are just two digits left for them: 0 and 5. Since RESTRAINT must not start with a zero, it follows that

$$R = 5, \qquad E = 0.$$

So, in summary:

E	D	L	T	N	R	I	M	A	S
0	1	2	3	4	5	6	7	8	9

and, in particular,

$$\text{RESTRAINT} \equiv 509358643.$$

Particular solution II

We are told

$$
\begin{array}{r}
\text{U N D E R} \\
+\text{W A T E R} \\
+\text{E R R O R} \\
\hline
=\text{D R O W N}
\end{array}
$$

We have[6]

$$3R = N + 10\alpha, \tag{1}$$

$$2E + \theta + \alpha = W + 10\beta, \tag{2}$$

$$D + T + R + \beta = \theta + 10\gamma, \tag{3}$$

$$N + A + \gamma = 10\delta, \tag{4}$$

$$U + W + E + \delta = D, \tag{5}$$

where α, β, γ, δ (the 'carries') are each 0, 1, or 2.

---

[6] From now on I write θ for the letter O (to avoid confusion with the numeral 0, zero).

It is clear from (4) that

$$\delta = 1, \tag{6}$$

and so from (5) (since U, W, E are different and non-zero) that

$$D \geq 7. \tag{7}$$

If $\beta = 2$ we would have by (2) that $E \geq 5$; and so by (5) and (6) that $E = 5, D = 9$. Hence we would have $\theta \leq 8$, and so $2E + \theta + \alpha \leq 20$. Since $W \geq 1$, it follows by (2) that

$$\beta \leq 1. \tag{8}$$

Temporarily adding (3) and (5) we also have (since T, R, U, W, E are all different, and so have a sum ≥ 10) that

$$\gamma = 1. \tag{9}$$

We note from (1) that

$$N \neq 0, \qquad N \neq 5. \tag{10}$$

We develop Table 8.2 by taking the three possible values of D; using (5), (6) we note the values that U, W, E (in some order) must have; and then consider the remaining digits (though not 0 or 5, by (10)) as possible values of N. For each, we deduce A and γ from (4)

Table 8.2

D	U, W, E	N	A	γ	R	α	T	θ	β	E	W	U
7	1, 2, 3	4	5	1	8	2	0	6	1	**		
		6	*									
		8	0	2	6	1	**					
		9		*								
8	1, 2, 4	3		*								
		6		*								
		7	*									
		9	0	1	3	0	5	6	0	**		
							6	7	0	**		
							5	7	1	2	1	4
9	1, 2, 5	3		*								
		4	*									
		6		*								
		7	*									
		8	0	2	6	1	7	3	1	**		
	1, 3, 4	2		*								
		6	2	2	*							
		7		*								
		8	0	2	6	1	7	2	0	**		

and (9), and R and α from (1). We then use (3) and (8) to determine what values (if any) T, θ, β may have (in the table I use ** to mean 'no valid pair of digits is available'); and we then use (2) to determine what values (if any) E and W may have: (5) then gives us U.

From this we have that there is just one possible set of the nine letters used in the addition sum, using all the digits except 6. So I (which does not occur in the addition sum, but which occurs in UNINTENDED) must be 6.

So, in summary:

A	W	E	R	U	T	I	θ	D	N
0	1	2	3	4	5	6	7	8	9

And, in particular,

$$\text{UNINTENDED} \equiv 4969529828.$$

Particular solution III

There are ten different capital letters used in the problem,[7] so all ten digits 0, 1, 2, . . . , 9 are represented.

We are told 'ALICE is odd', so

$$E \text{ is odd}; \tag{1}$$

and 'ALICE $-$ DNUθR $=$ RθUND', so

$$\begin{array}{r} R\ \theta\ U\ N\ D \\ +\ D\ N\ U\ \theta\ R \\ \hline = A\ L\ I\ C\ E \end{array}$$

We have

$$D + R = E + 10\alpha, \tag{2}$$

$$N + \theta + \alpha = C + 10\beta, \tag{3}$$

$$2U + \beta = I + 10\gamma, \tag{4}$$

$$\theta + N + \gamma = L + 10\delta, \tag{5}$$

[7] From now on I write θ for the letter O.

$$R+D+\delta=A, \tag{6}$$

where α, β, γ, δ (the 'carries') are each either 0 or 1.

From (2) and (6) we have that $A-E=10\alpha+\delta$; and so, since $A-E<10$ and $A\neq E$, we have

$$\alpha=0, \qquad \delta=1, \tag{7}$$

$$A=E+1. \tag{8}$$

β must be 0 or 1. We now show that no solution is possible when $\beta=1$.

Suppose that $\beta=1$. Then by (3), (5), (7), $L-C=\gamma$, and so, since $L\neq C$, we have that $\gamma=1$. (Recall that we now have $\alpha=0$, $\beta=\gamma=\delta=1$.) Now one of the ten capital letters must stand for 9. But:

by (2)	$D=9\rightarrow E>9$,	by (2)	$R=9\rightarrow E>9$,
by (3)	$C=9\rightarrow N+\theta=19$,	by (5)	$L=9\rightarrow\theta+N=18$,
by (4)	$U=9\rightarrow I=U$,	by (4)	$I=9\rightarrow U=I$,
by (5)	$N=9\rightarrow\theta=L$,	by (5)	$\theta=9\rightarrow N=L$,
by (8)	$E=9\rightarrow A=10$,	by (8)	$A=9\rightarrow E$ is even.

Consequently

$$\beta=0. \tag{9}$$

It follows immediately from (3), (5), (7) that $C-L=10-\gamma$, and, so, that

$$\gamma=1, \tag{10}$$
$$C=9, \qquad L=0. \tag{11}$$

Rewriting (2)–(6) using (7)–(11) we have

$$\left.\begin{array}{l} D+R=E, \\ N+\theta=9, \\ 2U=I+10, \\ A=E+1. \end{array}\right\} \tag{12}$$

(We could at this stage proceed simply by enumeration of cases, but a technique that avoids the need to do that is available to us: it is one that can sometimes be used to advantage in other problems in which there are ten different capital letters essentially involved in the Alphametric, so we'll use it here.)

The eight letters occurring in (12) stand for the digits $1, 2, \ldots, 8$ in some order, and so sum to 36. Using (12) to replace $(D + R)$, $(N + \theta)$, I, and A in that sum, we obtain $3U + 3E = 36$, so that

$$U + E = 12. \tag{13}$$

Since (by (11)) $I \neq 0$, we have (by (12)) that $2U > 10$; since (by (1)) E is odd, we have (by (13)) that U is odd; and since (by (11)) $U \neq 9$, it follows that

$$U = 7, \qquad E = 5. \tag{14}$$

(We then have by (12) that $I = 4$; that $A = 6$; and that D, R are 2, 3 in some order, and N, θ are 1, 8 in some order.)

It follows from (11) and (14) that

$$\text{CLUE} \equiv 9075.$$

Particular solution IV

We are told[8]

$$
\begin{array}{r}
\text{R I D E R} \\
+\,\text{D R}\,\theta\,\text{P S} \\
+\,\text{R E I N S} \\
+\,\text{H}\,\theta\,\text{R S E} \\
\hline
=\,\text{L}\,\theta\,\text{S E S}
\end{array}
$$

We have

$$R + S + E = 10\alpha, \tag{1}$$

$$P + N + S + \alpha = 10\beta, \tag{2}$$

$$D + \theta + I + R + \beta = S + 10\gamma, \tag{3}$$

$$I + R + E + \gamma = 10\delta, \tag{4}$$

$$2R + D + H + \delta = L, \tag{5}$$

where $\alpha, \beta, \gamma, \delta$ (the 'carries') are each 0, 1, 2, or 3.

We note that none of R, D, H, L can be zero.

---

[8] From now on I write θ for the letter O.

In this problem it seems best to concentrate first on obtaining the values of the carries: α, β, γ, δ.

If $\delta = 2$, we would have by (5) that $L = 9$, $R = 1$; and so by (4) that $I + E + \gamma = 19$. But we would also have that $I + E + \gamma \leq 8 + 7 + 3$. Hence $\delta \neq 2$. Clearly, by (4), δ can not be 0 or 3. It follows that

$$\delta = 1. \tag{6}$$

If we temporarily add the equations (2), (4), and (5), we obtain

$$P + N + S + I + 3R + E + D + H - L = 10\beta + 9 - \alpha - \gamma;$$

since different letters stand for different digits (and $R \neq 0$) the left-hand-side of this must be at least 21; and so $\beta \geq 2$. But by (2) we have that $\beta \leq 2$. So

$$\beta = 2. \tag{7}$$

By (5) and (6) we have that $R \leq 2$, and so by (1) that

$$\alpha = 1. \tag{8}$$

Subtracting (4) from (1) (and using (6), (8)) we have

$$S = I + \gamma. \tag{9}$$

It follows from (3) (using (7), (9)) that $D + \theta + R + 2 = 11\gamma$, so that (since $R \leq 2$) we have

$$\gamma = 1, \tag{10}$$

$$D + \theta + R = 9, \tag{11}$$

$$I + 1 = S. \tag{12}$$

Having determined the values of α, β, γ, δ, we can now rewrite (1)–(5): we have, in addition to (11), (12),

$$P + N + S = 19, \tag{13}$$

$$I + R + E = 9, \tag{14}$$

$$2R + D + H = L - 1, \tag{15}$$

By (15) we have that $L = 8$ or 9.

If $L = 8$ we would have that $R = 1$ and that D, $H = 2$, 3 (in some order). We would then have by (14) that $I + E = 8$. But neither I nor E could be 1, 2, or 3; and if either were 0, the other would equal L. So $L \neq 8$.

Consequently

$$L = 9. \tag{16}$$

By (15) we have that $R = 1$ or 2.

If $R = 1$ we would have that D, H $= 2, 4$ (in some order), and by (14) that $I + E = 8$. Bearing in mind that we've used up 1, 2, 4, 9, and keeping an eye on (12), it would follow that $I = 5$, $E = 3$, and $S = 6$. That would leave just 0, 7, 8 available for N, θ, P (in some order); and no two of these sum to 13, so that (13) could not be satisfied.

Consequently

$$R = 2. \tag{17}$$

By (15), (16), we then have that D, H $= 1, 3$ (in some order), and by (14) that $I + E = 7$. Bearing in mind that we've used up 1, 2, 3, and by (12) that $I \neq 0$, it follows that

$$E = 0, \qquad I = 7, \qquad S = 8. \tag{18}$$

This leaves just 4, 5, 6 available for N, θ, P (in some order), and so, by (13), (18),

$$N, P = 5, 6 \text{ (in some order)}, \tag{19}$$

$$\theta = 4. \tag{20}$$

It follows from (11), (17), and (20) that

$$D = 3, \tag{21}$$

and so from (15), (16), (17), and (21) that

$$H = 1. \tag{22}$$

We have determined the values of eight of the capital letters, but there is no information that will enable us to resolve the ambiguity (19): we have

E	H	R	D	θ	N P	P N	I	S	L
0	1	2	3	4	5	6	7	8	9

We do, however, have enough information to determine the numerical value of HθRSE:

$$H\theta RSE \equiv 14280.$$

Particular solution V

We are told

```
    T R U S T
+   H O U S E
+   F O R T E
─────────────
= H O T E L S
```

We have[9]

$$T + 2E = S + 10\alpha, \tag{1}$$

$$2S + T + \alpha = L + 10\beta, \tag{2}$$

$$2U + R + \beta = E + 10\gamma, \tag{3}$$

$$R + 2\theta + \gamma = T + 10\delta, \tag{4}$$

$$T + F + \delta = \theta + 9H, \tag{5}$$

where α, β, γ, δ (the 'carries') are each 0, 1, or 2.

We note that none of T, H, F can be zero.

Eliminating θ between (4) and (5), we have

$$\begin{aligned} 18H &= 2F + R + T + \gamma - 8\delta \\ &\leq (2 \times 9) + 8 + 7 + 2 \\ &= 35. \end{aligned}$$

Hence

$$H = 1. \tag{6}$$

In this problem I now see no significantly better course than to go straight into 'enumeration of cases'. The work is straightforward, but very long (the full table occupies about 150 lines); to include it here would be to give it a spurious importance. So I confine myself giving the eventual result.[10]

---

[9] From now on I write θ for the letter O.
[10] But see Appendix.

We at length obtain the unique result

S	H	N	θ	U	F	T	E	L	R
0	1	2	3	4	5	6	7	8	9

and, in particular,

$$\text{RESTFULNESS} \equiv 97065482700$$

Composer's problem

My wife wanted a problem for the Christmas issue of the local riding club's newsletter—'Not too difficult: how about an Alphametic?' I had of course seen and solved Alphametics (the earliest that I remember is the classic A + MERRY + XMAS = TURKEY), but I had not previously thought of composing one; 'Reindrop' (Problem IV) took me longer than I care to recall. I was rather pleased with it: the words seemed appropriate to the occasion, and (as I only later discovered) its solution is straightforward, involving remarkably little in the way of enumeration of cases.[11] But it does have a flaw, in that it does not have a completely unique solution (N and P are 5 and 6, but there is nothing to tell us which is which).

My subsequent endeavours to produce Alphametics that did not have the 'Reindrop' flaw[12] (and with sensible words that formed sensible messages) produced 'Underwater Error' (Problem II); and then 'Stand Still, Damnit' (Problem I)—which had an additional twist to it: since none of the eight letters in the sum represents zero, I could ask for the values of *two* further letters (R and E in RESTRAINT).

All these first four problems are capable of fairly short solution (with little or no 'enumeration of cases'); and in the solutions that I have given I have tried to illustrate some of the ploys that may be useful in dealing with other Alphametics. But THF (Problem V) seems to have no short solution: apart from the relatively immediate

---

[11] Euphemism for 'trial and error'.

[12] 'Alice' (Problem III) was *not* part of these endeavours! 'Alice' started simply as the question to myself: 'If one adds a number to its reversal, can the whole sum involve all ten digits?' So restraints had to be added (e.g. 'ALICE is odd')—and even so, 'Alice' and 'Reindrop' have the same flaw. (My device of asking just for certain letters—C, L, U, E in the case of 'Alice'—does not remove the flaw: it merely disguises it.)

deduction that $H = 1$ (rather than 2), its solution involves tenacity rather than ingenuity.[13] But, at least, it does have a unique solution (and I cannot find any other company name that has this unusual alphametic property!)

I do not know how other people compose Alphametics: indeed, I am not at all sure how I do it myself. But of one thing I am sure: they are even more fun to compose than to solve.

Appendix : Detailed solution of Problem V

We adopt Particular Solution V up to equation (1).

We develop Table 8.3 by taking all possible pairs of values of T and E; then, using (1), we deduce S and α; then, using (2), we deduce L and β; then, using all possible remaining values of θ, we deduce, using (5), F and δ; then, using (4), R and γ; finally, using (3), U.

In each line we stop if and when we arrive at a duplication (*).

Table 8.3

T	E	S	α	L	β	θ	F	δ	R	γ	U
	0	*									
	3	8	0	*							
	4	0	1	3	0	≥5	*				
	5	*									
	6	4	1	*							
						0	*				
2	7	6	1	5	1	3	8	2	*		
							9	1	*		
						4	9	2	*		
						≥8	*				
	8	*									
	9	0	2	4	0	3	8	2	*		
						≥5	*				

[13] I attach an Appendix giving the long and boring demonstration that the answer to THF is unique.

Table 8.3—*continued*

T	E	S	α	L	β	θ	F	δ	R	γ	U
3	0	*									
	2	7	0	*							
	4	*									
	5	*									
	6	5	1	4	1	0	*				
						2	7	1	8	1	*
							8	0	9	0	*
									*		
						≥7					
	7	*									
	8	9	1	2	2	0	4	2	*		
							5	1	*		
							6	0	*		
							*				
						≥4					
	9	*									
4	0	*									
	2	8	0	0	2	3	6	2	*		
							7	1	6	2	*
						5	9	1	3	1	*
						6	9	2	*		
						≥7	*				
	3	0	1	5	0	2	6	1	8	2	*
							7	0	9	1	*
									*		
						6	9	2	*		
						≥7	*				
	5	*									
	6	*									
	7	8	1	*							
	8	0	2	6	0	2	5	2	*		
							7	0	*		
						3	7	1	*		
						5	9	1	2	2	*
									3	1	*
						≥7	*				
	9	2	2	0	1	3	6	2	*		
							7	1	6	2	*
							8	0	8	0	*
									*		
						5	8	2	*		
						≥6	*				

Table 8.3—*continued*

T	E	S	α	L	β	θ	F	δ	R	γ	U
5	0	*									
	2	9	0	3	2	0	*				
						4	6	2	*		
							7	1	6	1	*
							8	0	*		
						6	8	2	*		
						≧7	*				
	4	3	1	2	1	0	*				
						6	8	2	*		
							9	1	*		
						7	9	2	*		
						≧8	*				
	6	7	1	0	2	2	4	2	*		
						3	*				
						4	8	0	*		
						≧8					
	7	9	1	4	2	0	2	2	*		
							3	1	*		
						2	6	0	0	1	*
						3	6	1	8	1	*
						6	8	2	*		
						8	*				
	8	2	2	*							
	9	3	2	*							
6	0	*									
	2	0	1	7	0	3	4	2	*		
							5	1	8	2	*
									9	1	*
						4	5	2	*		
						5	8	0	*		
						8	9	2	*		
						9	*				
	3	2	1	*							
	4	*									
	5	*									
	7	0	2	8	0	2	3	2	*		
							4	1	*		
							5	0	*		
						3	4	2	*		
							5	1	9	1	4 ←¶
						4	5	2	*		
						5	*				
						9	*				
	8	2	2	*							
	9	4	2	*							

Table 8.3—*continued*

T	E	S	α	L	β	θ	F	δ	R	γ	U
7	0	*									
	2	*									
	3	*									
	4	5	1	8	1	0	2	0	6	1	*
						2	3	1	*		
						3	*				
						6	*				
						9	*				
	5	*									
	6	9	1	*							
	8	3	2	5	1	0	2	0	*		
						2	4	0	*		
						4	6	0	*		
						5	*				
						9	*				
	9	5	2	*							
8	0	*									
	2	*									
	3	4	1	7	1	0	*				
						2	*				
						5	6	0	*		
						6	*				
						9	*				
	4	6	1	*							
	5	*									
	6	0	2	*							
	7	2	2	4	1	0	*				
						3	*				
						5	6	0	*		
						6	5	2	*		
						9	*				
	9	6	2	2	2	0	*				
						3	4	0	0	2	*
						4	3	2	*		
							5	0	0	0	*
						5	4	2	*		
						7	*				

Table 8.3—*continued*

T	E	S	α	L	β	θ	F	δ	R	γ	U
9	0	*									
	2	3	1	6	1	0	*				
						4	*				
						5	4	1	7	2	*
									8	1	*
						7	5	2	*		
						8	7	1	*		
	3	5	1	0	2	2	*				
						4	2	2	*		
						6	4	2	*		
						7	6	1	4	1	*
						8	6	2	*		
							7	1	2	1	*
	4	7	1	*							
	5	*									
	6	*									
	7	3	2	*							
	8	5	2	*							

By this exhaustive (and somewhat exhausting) process we eventually have that there is a unique solution: namely $H = 1$, $T = 6$, $E = 7$, $S = 0$, $L = 8$, $\theta = 3$, $F = 5$, $R = 9$, $U = 4$; and, finally, the previously unmentioned letter N (in RESTFULNESS) must be the one remaining digit (2): $N = 2$.

9

UNCOMMON BIRTHDAYS

Problem

'There are quite a number of people in that room,' said the Pieman, 'all random visitors, and all I know about them is that none of them was born on 29 February. Now tell me: what are the chances that no two of them have the same birthday? Same month and day of the month, that is; they don't have to be the same year. Once chance in . . . what, would you say?'

'It's obvious,' said Simple Simon. 'One chance in the number of people that there are in the room—to the nearest whole number, at any rate.'

'I don't think you quite understood me . . . ,' started the Pieman.

'Oh yes I did,' said Simple Simon. 'You count them and see if I'm wrong.'

He wasn't wrong.

So how many people were in the room?

Solution

If there are two people in a group, the chance of their not having the same birthday is $\left(\frac{364}{365}\right)$.[1] Introduce a third person to the group: there are 363 days still available for his birthday not to coincide with either of the first two; so the chance of no two or the three having the same birthday is $\left(\frac{364}{365}\right)\left(\frac{363}{365}\right)$. Introduce a fourth: the chance becomes $\left(\frac{364}{365}\right)\left(\frac{363}{365}\right)\left(\frac{362}{365}\right)$. Keep going: with n people in the group, the chance of no two of them having the same birthday is

$$C_n = \left(\frac{364}{365}\right)\left(\frac{363}{365}\right)\cdots\left(\frac{366-n}{365}\right). \tag{1}$$

With a pocket-calculator, the calculation and tabulation of C_n (for increasing values of n) is straightforward and rapid; lacking a pocket-calculator, but having a table of logarithms, it is straightforward but rather tedious; lacking even a table of logarithms, it is practicable only for $n < N$, where N is a number that depends solely on one's tenacity (euphemism for pigheadedness).

So we tabulate.

Probably the easiest way is to calculate successive values of $1/C_n$, using

$$\frac{1}{C_n} = \left(\frac{365}{366-n}\right)\left(\frac{1}{C_{n-1}}\right).$$

To start with,

n	$1/C_n \cong$
2	1·002747
3	1·008272
4	1·016628
. . .	. . .

Later we come to

n	$1/C_n \cong$
. . .	. . .
22	1·907288
23	2·029621
. . .	. . .

[1] Leap years are an undesirable complication. So, for simplicity, I assume that nobody, ever, is born on 29 February. (My apologies to the 1/1461 of the population who are.)

which confirms a classic result:[2]

> *If you know 23 or more people's birthdays, it is more likely than not that two of them are the same (as to day and month).*

We go on, through

n	$1/C_n \cong$
...	...
30	3·405023
...	...
40	9·193864
...	...
50	33·75366
...	...

to

n	$1/C_n \cong$
...	...
52	45·46386
53	53·01697
54	62·02305

which gives us the answer to the problem: there are 53 people in the room.

Composer's problem

This all started in a restaurant. A girl at the next table (who, I think, had surreptitiously counted the number of people dining—there were about 25 of us) put a question to her companions: 'I'll bet that there

---

[2]   My copy of Rouse Ball (W. W. Rouse Ball, *Mathematical Recreations and Essays*, 11th edn. (Macmillan, 1944)) has (p. 45):

> *Here is a result (due to H. Davenport) which many people find surprising. If you know more than 23 people's birthdays, it is more likely than not that two of them are the same (as to day and month).*

I have amended Rouse Ball's 'more than 23' to '23 or more' in the above text. (See 'Composer's Problem'.)

are at least two people dining here who have the same birthday. Any takers?'

I won't give the full detail of their subsequent conversation (to which I listened unashamedly); but two things emerged from it: first, that one of the party thought it 'obvious' that the chance of no two people having the same birthday must be one in the-number-of-people-in-the-room (they had quite a time convincing him that he was in error): and, second, that the girl was sure that the break-point was 24 (i.e. 'If there are 24 or more people in the room then it's more likely than not that two of them have the same birthday') rather than 23.

Rather late that night, I looked at my 40-year-old copy of Rouse Ball, and unearthed my pocket-calculator.[3]

Four things emerged.

1 The girl's insistance on 24 (rather than 23) was in complete accord with my copy of Rouse Ball ('more than 23'); but my pocket-calculator told me—repeatedly—that it was 23 ('23 or more').[4]

2 The man who had said 'one in-the-number-of-people-in-the-room' was clearly in error—but he was in error in one direction when the number was small, and in error in the other direction when the number was large. So where was the cross-over?

Finding out was just a matter of extending the table. By the time that n is over 50, $1/C_n$ is advancing by 7 or 8 for each increase of 1 in n: so the actual result that $1/C_{53} \cong 53 \cdot 0$ is a piece of luck. (It might have turned out that $1/C_{52} \cong 49$, $1/C_{53} \cong 57$, or some such: an uninteresting cross-over.) It was only then that it occurred to me that I had a possible problem—and I had my anonymous dinner neighbour to thank for the context in which to set it.

3 With the bit between my teeth, I continued with the table—mainly to see if anything else of interest turned up.

I did not find anything—but you may: selected entries are given in Table 9.1.

---

[3] I don't like using calculators in *setting* problems. But here I wasn't setting a problem: I was merely verifying a result. At least, that's what I told myself at the time.

[4] I first heard of the problem in about 1937, and was told that the answer was '23 or more'. (There were no calculators, and I had not then met logarithms, so I did *not* try to verify the result!) But it was not until this 1987 incident that I realized that my copy of Rouse Ball said 'more than 23' rather than '23 or more'. It is unthinkable that Davenport and Rouse Ball did not have the correct answer in the first place. So when and where did the printed error creep in? I haven't found out, yet.

Table 9.1

n	$1/C_n \cong$	n	$1/C_n \cong$
1	1	200	$6 \cdot 210696 \times 10^{29}$
20	$1 \cdot 699057$	220	$1 \cdot 623189 \times 10^{37}$
40	$9 \cdot 193864$	240	$6 \cdot 688337 \times 10^{45}$
60	$1 \cdot 701450 \times 10^2$	260	$6 \cdot 766821 \times 10^{55}$
80	$1 \cdot 167296 \times 10^4$	280	$3 \cdot 104987 \times 10^{67}$
100	$3 \cdot 254690 \times 10^6$	300	$1 \cdot 601196 \times 10^{81}$
120	$4 \cdot 099793 \times 10^9$	320	$4 \cdot 090565 \times 10^{97}$
140	$2 \cdot 638797 \times 10^{13}$	340	$9 \cdot 342744 \times 10^{117}$
160	$1 \cdot 003296 \times 10^{18}$	360	$1 \cdot 273114 \times 10^{146}$
180	$2 \cdot 679986 \times 10^{23}$	365	$6 \cdot 873064 \times 10^{156}$

(Any intermediate value can be calculated from the nearest entry in the table, with—at most—ten multiplications and ten divisions, using $365C_n = (366-n)C_{n-1}$.)

4   I was a little bit worried about the accuracy of my results: were accumulated rounding errors affecting things too much?

So it seemed to be a good idea to calculate $1/C_{365}$ by a different method, as a check on the accuracy of the last entry in the table: I used Stirling's approximation[5] for $\Gamma(365)$.

Stirling's asymptotic formula for the Gamma-function is:

$$\Gamma(x) = e^{-x} x^{x-\frac{1}{2}} (2\pi)^{\frac{1}{2}} \nabla 1 + \frac{1}{12x} + \frac{1}{288x^2} + \dots \bigg). \qquad (2)$$

So

$$C_{365} = 364! \div 365^{364}$$

$$= \Gamma(365) \div 365^{364}$$

$$\cong e^{-365} \, 365^{\frac{1}{2}} \, (2\pi)^{\frac{1}{2}} \left(1 + \frac{1}{12 \times 365} + \frac{1}{288 \times 365^2} \right)$$

$$\cong e^{-365} \times 47 \cdot 90000076. \quad [6]$$

So

$$1/C_{365} \cong 6 \cdot 873064 \times 10^{156},$$

(which satisfactorily confirms the accuracy of the results in the earlier table).

---

[5] See e.g. E. T. Whittaker and G. N. Watson, *Modern Analysis*, 4th edn. (Cambridge University Press, 1946), §12.33.

[6] I include explicit mention of this step only because of the curious occurrence of all those zeros.

The original problem is, of course, trivial: solving it requires only the ability to multiply and divide, and either inordinate tenacity or a calculator. My only excuse for including it is that I enjoyed the context in which I could set it.

10

FOCAL LENGTHS

Problem I

There are four large trees in the plain: an Oak, a Pine, a Quince, and a Rowan. Ceiling Church lies directly between the Pine and the Rowan, and directly between the Oak and the Quince.[1]

The Curate of Ceiling—keen and fit, like most curates—recommends two walks in the neighbourhood. Each starts at the church, and each finishes at the church: but one visits in turn the Oak, the Pine, and the Quince; and the other visits in turn the Rowan, the Quince, and the Pine. (They are along straight paths—apart, of course, from turning the corners at the trees.) The two walks are of the same total length.

The Oak and the Quince are further apart than are the Pine and the Rowan.

So which of the Oak and the Rowan is nearer to Ceiling Church?

---

[1] This is a repetition of (some of) the information given in Chapter 3 of *Mathematical Byways in Ayling, Beeling, and Ceiling*. The current problem is independent of the earlier one.

Problem II

There are survey points at Failing and Feeling, which are 25 miles apart. From each of them Tom (as part of the practical work for his Diploma in Advanced Surveying) made observations of the church spires at Gleaming and Glowing. He found that the distance of Gleaming from Failing was 4 miles less than the distance of Glowing from Feeling; that the distance of Glowing from Failing was 4 miles less than the distance of Gleaming from Feeling; that the distance of Gleaming from Feeling was 7 miles less than the distance of Gleaming from Failing; and that the church spire at Healing stood both between Failing and Gleaming and between Feeling and Glowing.

'What', asked his examiner, 'is the difference in the distances from Healing to Gleaming and from Healing to Glowing?'

Particular solution I[2]

We have (see Diagram 10.1) two triangles OPQ, PQR on the same side of the same base (PQ). We are told that OQ, PR met internally at C; that

$$OP + OQ = PR + QR;\qquad(1)$$

and that

$$OQ > PR.\qquad(2)$$

We want to find out which of CO, CR is the greater.

By (2) there is a point E on OQ (see Diagram 10.2), internal to OQ, such that

$$EO = PR.\qquad(3)$$

Then (by (1), (3))

$$OP = QR - (OQ - OE)$$

so (E is internal to OQ)

$$OP = QR - QE$$

so (3 sides of Δ)

$$OP < RE.\qquad(4)$$

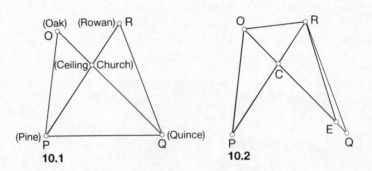

Diagrams 10.1–10.2

---

[2] I *know* that everywhere else I look at the General Solution before I go on to the Particular Solutions. But the General Solution is going to be quantitative and trigonometrical; and this particular problem is qualitative and has a simple solution in pure geometry: it deserves a separate presentation. (And, anyhow, the problem is not really mine: it is a slight generalization of what I always think of as 'Robson's Problem' (see the 'Composer's Problem' section).)

We now consider the two triangles EOR and PRO. We have

(by (3))	EO = PR,	
(common)	OR = RO,	
(by (4)	RE > OP;	

so

$$\widehat{EOR} > \widehat{PRO}.$$

That is:

$$\widehat{COR} > \widehat{CRO}.$$

Hence (we are now looking at the triangle COR)

$$CR > CO.$$

General solution

1   We are given four points F_1, F_2, G_1, G_2, and told that

$$F_1G_1 + F_2G_1 = F_1G_2 + F_2G_2, \tag{5}$$

and that F_1G_1, F_2G_2 meet internally (let them do so at H).

To fix our ideas, we take rectangular Cartesian coordinates with origin at the mid-point of F_1F_2 and with F_1, F_2 on the X-axis: at $(-c, 0)$ and $(c, 0)$, say. Then there exist a, b, γ, δ such that G_1 is at $(a \cos \gamma, b \sin \gamma)$ and G_2 is at $(a \cos \delta, b \sin \delta)$,[3] where

$$a^2 = b^2 + c^2. \tag{6}$$

(See Diagram 10.3.) Noting that

$$(c + a \cos \theta)^2 + (b \sin \theta)^2 = (a + c \cos \theta)^2,$$

we obtain

$$F_1G_1 = a + c \cos \gamma, \tag{7a}$$

$$F_2G_1 = a - c \cos \gamma, \tag{7b}$$

$$F_1G_2 = a + c \cos \delta, \tag{7c}$$

$$F_2G_2 = a - c \cos \delta. \tag{7d}$$

[3] This is not immediately obvious. (It's obvious only if you know about the coordinate geometry of ellipses: in that case (5) has told you straight away that G_1 and G_2 are points on an ellipse whose foci are F_1 and F_2, and everything up to (7) follows automatically. But let's suppose that you *don't* know about ellipses.)
We can certainly say that there exist a, b, γ such that G_1 is at $(a \cos \gamma, b \sin \gamma)$, where $a^2 = b^2 + c^2$; and we can certainly say that there exist a', b', δ such that G_2 is at $(a' \cos \delta, b' \sin \delta)$, where $a'^2 = b'^2 + c^2$. We then find that $F_1G_1 + F_2G_1 = 2a$, and that $F_1G_2 + F_2G_2 = 2a'$. We then have by (5) that $a = a'$ (and, so, that $b = b'$).

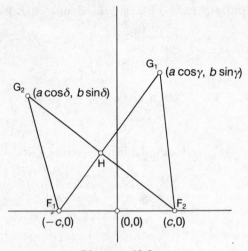

Diagram 10.3

2   For the moment we take λ, μ such that

$$G_1H = \lambda F_1G_1, \tag{8a}$$

$$G_2H = \mu F_2G_2. \tag{8b}$$

Then the coordinates of H are (expressed in two different ways)

$$-\lambda c + (1-\lambda)a \cos \delta, \qquad (1-\lambda)b \sin \delta \tag{9a}$$

and

$$\mu c + (1-\mu)a \cos \gamma, \qquad (1-\mu)b \sin \gamma. \tag{9b}$$

Equating the two X-coordinates, and equating the two Y-coordinates, we have from (9) two simultaneous equations in λ and μ, which we solve to get

$$\lambda = \frac{\tan \phi(a \cos \phi - c \cos \theta)}{a \sin \phi + c \sin \theta} \tag{10a}$$

$$\mu = \frac{\tan \phi(a \cos \phi + c \cos \theta)}{a \sin \phi + c \sin \theta} \tag{10b}$$

where

$$\theta = \tfrac{1}{2}(\delta + \gamma), \tag{11a}$$

$$\phi = \tfrac{1}{2}(\delta - \gamma). \tag{11b}$$

3 From (7), (8), (10), (11) we then have

$$G_1H = \frac{\tan\phi(a\cos\phi - c\cos\theta)\{a + c\cos(\theta - \phi)\}}{a\sin\phi + c\sin\theta}, \quad (12a)$$

$$G_2H = \frac{\tan\phi(a\cos\phi + c\cos\theta)\{a - c\cos(\theta + \phi)\}}{a\sin\phi + c\sin\theta}, \quad (12b)$$

On subtraction, a gratifying simplification occurs:[4]

$$G_2H - G_1H = 2c\tan\phi\sin\phi\cos\theta. \quad (13)$$

Now we know by (7), (11) that

$$F_1G_1 - F_2G_2 = 2c\cos\phi\cos\theta,$$

and so by (13)[5]

$$G_2H - G_1H = (F_1G_1 - F_2G_2)\tan^2\phi. \quad (14)$$

4 Suppose that we are given three quantities c, g, h, and told that

$$F_1F_2 = 2c, \quad (15a)$$

$$F_1G_1 - F_2G_1 = g, \quad (15b)$$

$$\left.\begin{array}{l} F_1G_1 - F_2G_2 \\ F_1G_2 - F_2G_1 \end{array}\right\} = h. \quad (15c)$$

Then by (7), (11b) we can successively calculate γ, δ, ϕ, and, so, $G_2H - G_1H$. But the explicit expression for $G_2H - G_1H$ in terms of c, g, h is rather ugly: I do not reproduce it here.

Particular solution II

Labelling Failing and Feeling as F_1, F_2, Gleaming and Glowing as G_1, G_2, and Healing as H, we have that

$$F_1F_2 = 25 \text{ miles},$$

$$F_1G_1 - F_2G_1 = 7 \text{ miles},$$

$$\left.\begin{array}{l} F_1G_1 - F_2G_2 \\ F_1G_2 - F_2G_1 \end{array}\right\} = -4 \text{ miles};$$

and that F_1G_1, F_2G_2 meet internally at H.

[4] Should we have seen it coming? Perhaps there's a better way of getting to (13) (or (14)) than the one that I've taken. But I haven't found it—yet.
[5] (14) provides us with an alternative way of arriving at the answer to Problem I.

Then by (7) we have

$$\cos \gamma = 7/25, \qquad \cos \gamma + \cos \delta = -8/25;$$

and so

$$\tan \tfrac{1}{2}\gamma = 3/4, \qquad \tan \tfrac{1}{2}\delta = 2.$$

Hence by (11)

$$\tan \phi = 1/2,$$

and so by (14)

$$G_1 H - G_2 H = 1 \text{ mile}.$$

Two examples of possible configurations are given in Diagram 10.4.

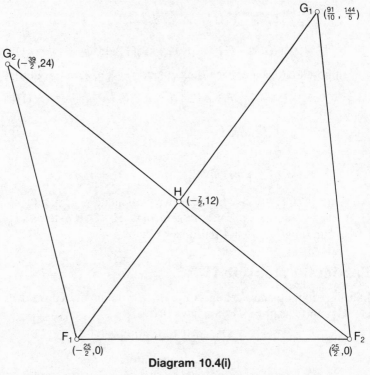

Diagram 10.4(i)

In each case (and in an infinity of other cases) $F_1 F_2 = 25$, $F_1 G_1 - F_2 G_1 = 7$, $F_2 G_2 - F_1 G_1 = F_2 G_1 - F_1 G_2 = 4$; and so $G_1 H - G_2 H = 1$.

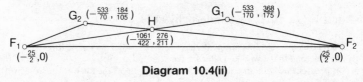

Diagram 10.4(ii)

Composer's problem

1 It was at some time in the late 1920s or the early 1930s that Mr A. Robson, then Senior Mathematics Master at Marlborough College, set a problem:

> PQR is an isosceles triangle with PR = QR; OPQ is a triangle having the same perimeter, and is such that OQ meets PR internally (in C, say). Show by methods of Pure Geometry that CR > CO.

The problem was still being set at Marlborough in 1951[6].

I can plead in mitigation of my plagiarism only that Problem I is a slight generalization—in that it does not restrict PQR to be isosceles.

2 The Analytic Geometry approach can be a little fearsome unless one recognizes the elliptic property at an early stage, to get (7). (And making the change (11) helps a bit, too.)

3 One interesting thing about the result stated in Section 4 of the General Solution is that neither a nor b occurs in it. That fact allowed me to set a problem that at first sight might be thought to have insufficient data.[7]

> With the coordinate system of Section 1 of the General Solution, and given constants c, g, h defined by (15), there are an infinity of possible positions for H: they lie on the hyperbola[8]
>
> $$\frac{x^2}{c^2 \cos^2 \alpha} - \frac{y^2}{c^2 \sin^2 \alpha} = 1,$$
>
> where
>
> $$\cos \alpha = \frac{\cos \theta}{\cos \phi} = \frac{1 + \cos(\gamma + \delta)}{\cos \gamma + \cos \delta},$$
>
> where, by (7),
>
> $$\left. \begin{array}{l} \cos \gamma = g/2c, \\ \cos \delta = (2h - g)/2c. \end{array} \right\}$$
>
> But for all such H we have by (14) that
>
> $$G_2H - G_1H = h \tan^2 \phi = c(2 \cos \alpha - \cos \gamma - \cos \delta): \text{ a constant.}$$

---

[6] I was a temporary probationary assistant mathematics master at Marlborough in 1951–2. The problem was new to me: *amour propre* dictated that I solve it before the next morning's classes. That was a late night.

[7] A situation virtually irresistible to a problemist.

[8] This very nearly became the basis for another problem. But I cannot think of one in which the interest outweighs the complications.

4   The next step was to find suitable numbers for insertion in the data to provide a simple answer (to Problem II).

Writing

$$\left.\begin{aligned} \tan \tfrac{1}{2}\gamma &= p/q, \\ \tan \tfrac{1}{2}\delta &= r/s, \end{aligned}\right\}$$

I had,[9] from (7), (11), (15),

$$2c = (q^2 + p^2)(s^2 + r^2)(qs + pr)/t,$$

$$g = (q^2 - p^2)(s^2 + r^2)(qs + pr)/t,$$

$$h = (qs - pr)(qs + pr)^2/t,$$

$$G_2H - G_1H = (qs - pr)(qr - ps)^2/t,$$

(where t is going to be my eventual 'dividing through' factor).

I then gave p, q, r, s (and, subsequently, t) various integer values, and eventually settled on a set $(3, 4, 2, 1)$ (and, subsequently, $t = 50$) that provided an integer solution having the property that

$$|G_2H - G_1H| = 1.$$

Extension

One problem suggests another. I had done what I set out to do (in Section 3 of the 'Composer's Problem'), but I was left with the nagging question: what *other* sets of values of p, q, r, s, t could I have chosen that would lead to integer values for $2c, g, h$ and have $|G_2H - G_1H| = 1$?

I could, of course, find several others by trial and error. But that is uninteresting.

I could take an obvious first step: there must exist an integer k such that

$$\frac{s}{r} = \frac{kq - p}{kp + q}$$

(from (17c, d)).

But at that stage I came to a grinding halt.

So my Extension Question is: Find the general parametric solution of (17) that yields all—and only—situations in which $2c, g, h$ are integers and $|G_2H - G_1H| = 1$.

---

[9]   Using a little foresight! (But only a little.)

11

PAR FOR THE COURSE

One of the difficulties about setting a problem in a golfing context is that the game is permeated with technical terms to an extent rivalled only by real tennis (and the meanings of those technical terms differ from country to country and from decade to decade—consider the word 'bogey', for example). So before I can even state the problem I need to give some definitions.

Each *hole* of a golf course has an associated *par* score. If a golfer *does a hole in par* he is usually quite pleased. If a golfer does a hole in one less than par he has achieved a *birdie* (and is very pleased). If he does a hole in one more than par he has achieved only a *bogey* (and is not so pleased). (That is all that you need to know about this part of scoring at golf: the arcana of eagles, albatrosses, double bogeys, and so on do not enter into the problem.)

In a *four-ball* match, a partnership of two players opposes a partnership of two other players. Each of the four players plays (his own ball at) each hole. On the completion of each hole the player who has done best at that hole scores a point for his partnership (irrespective of how his partner has fared). If players from opposing partnerships share 'best at that hole' (irrespective of how their partners have fared) then no point is awarded. As might be expected, the eventual winners of the match are the partnership with most points after the predetermined number of holes has been played.

Handicaps? As far as this problem is concerned, there aren't any. *My* golfers count the number of strokes that they have actually played (*not* some fictitious number of strokes that they might have played had they been better players).

Now that we've got that over . . .

Problem

Dick and Harry are golfers of exceptional consistency: neither of them has ever scored anything other than par at any hole—and in all probability neither of them ever will. Anne is neither as good nor as consistent: in a typical completed round (of 18 holes) she scores nine bogies, six pars, and three birdies. Precisely the same can be said of Belinda as has just been said of Anne.

'It would be a fair match,' mused Dick, 'if Anne and I took on Harry and Belinda at a four-ball. How about it?'

'Fair—but not very interesting,' said Anne. 'I'd rather have a less fair match. Belinda and I will challenge you and Harry, Dick. And, while we're about it, let's make it over a double round—36 holes instead of 18. Will you take us on?'

Belinda was making faint protesting noises in the background.

'Take you on?' said Dick. 'We'll beat you out of sight. I go round 18 holes in six strokes less than you; Harry goes round in six strokes less than Belinda; I reckon that we won't have to bother to play any of the holes after the 25th at all.'

'We'll see,' said Anne.

They played Anne's proposed four-ball match. Assuming that each played in accord with his or her established form, what was the result?

Solution

At each and every hole Dick and Harry get a par.

At half the holes Anne gets a bogey, and at half the holes Belinda gets a bogey; so the chances are that at a quarter of the holes both Anne and Belinda get bogies: that is, that at 9 holes out of the 36 the Dick–Harry partnership will win.

At $\frac{5}{6}$ of the holes Anne gets a par (or worse), and at $\frac{5}{6}$ of the holes Belinda gets a par (or worse): so the chances are that at $\frac{25}{36}$ of the holes both of them get par (or worse). That means that at $\frac{11}{36}$ of the holes one (at least) of them will get a birdie. That is, that at 11 holes out of the 36 the Anne/Belinda partnership will win.

So: Anne and Belinda win by 11 points to 9 over the 36-hole match.

Composer's problem

I was watching excerpts from the 1987 Ryder Cup on television. After the first group of foursomes Europe and the United States were all-square (2-all). Then came the first group of four-balls. The television commentator remarked with mild surprise on the selection of the European team; steadiness and consistency were the strengths that really mattered in four-balls, he felt; and, he also felt, the European captain did not seem to have borne this sufficiently in mind. The United States captain had. Europe won all four of that group of four-balls. I did not, however, see them do it: I was doing sums on the back of an envelope.

Steadiness and consistency are in fact the last things that one wants in a four-ball. What one wants is a pair of erratic geniuses—each of whom will produce a scatteration of birdies, eagles, bogies, and double bogies (in a round in which, if everything he did had to be counted, he might well lose badly to a steady consistent opponent). But in a scoring system in which his brilliancies matter a lot, and his wildnesses matter only if his (equally erratic) partner is wild at the same hole, he and his partner will come out on top.

The problem that I have set is perhaps somewhat artificial—and is certainly trivial. But it makes a point that some golf team captains (and certainly some television commentators) don't seem to realize.

12

CUBE THROUGH CUBE

Problem

'I have here', said Anne, 'a cube. Look at it. A perfectly normal solid cube. Incidentally, each of its face-areas is exactly 36 square inches.'

'You mean', said Tom, 'that each edge is 6 inches long.'

'Yes,' said Anne, 'but I'd like to talk about it in areas rather than in lengths, if you don't mind.'

'Fair enough,' said Dick; 'you're the one setting the problem.'

'Right,' said Anne. 'Now I want you to make a hole in it. As large a hole as you need—but it must be a genuine hole.'[1]

'Seems a pity to spoil it,' said Harry. 'What's the point?'

'When you've made the hole,' said Anne, 'you have to be able to pass one of these other cubes through it. here they are: four of them. They've got face-areas of 35, 38, 39, and 40 square inches. Which is the largest one that you can get through?'

'It's obvious,' said Tom: 'the 35 square inch one. I mean, it stands to reason that you can't pass a larger cube through a smaller one. Oh! Wait a minute . . .'

Tom eventually plumped for the 38-square-inch cube, Dick for the 39-square-inch one, and Harry for the 40-square-inch one. Why did they?[2] And who was correct?

---

[1] I won't go into the topological subtleties of 'when is a hole not a hole?' But, just to make sure that we're all thinking about the same thing: **B** has two holes in it; each of **A**, **D**, **O**, **P**, **Q**, **R** has one hole in it; and none of the other capital letters in the alphabet has a hole in it. (A 'hole' dug in the ground is usually not a hole at all—in the topological sense.)

[2] This is an unfair question! (Who can possibly tell their reasoning in arriving at—in two cases—incorrect results?) (But their arguments are quite persuasive—to them at least—at first sight.)

Solution

1   We start with a cube **C** (with edge-length $2n$, say; and, so, with face-area $4n^2$). Through it we intend to make a hole through which a cube **D** (with edge-length $2\mu n$; and, so, face-area $4\mu^2 n^2$) can be passed.

2   Tom, Dick, and Harry all want to find out whether μ can be sufficiently large for μ^2 to be greater than $\frac{38}{36}$, or $\frac{39}{36}$, or $\frac{40}{36}$.

There are holes that curve; there are holes that corkscrew (even though they may have straight axes); there are holes with all sorts of other peculiarities. But Tom, Dick, and Harry all assume that the holes that *they* are going to make are simple ones—with straight axes, and which, when viewed along their axes, appear as squares (of side $2\mu n$).

3   All three of them then make a much less reasonable assumption, arguing:

> If I look at **C** along a line through two of its opposite face-centres, all I see is a square of side $2n$. So with that direction as axis I can't make a hole that will accommodate **D** for any $\mu \geq 1$. I must twizzle **C**. If I do so about a line through two of its opposite face-centres (not the line that I'm looking along, of course: one of the other two), what I see will increase from a square of side $2n$ to a rectangle of sides $2n$, $(2\sqrt{2})n$ (when I've turned **C** through 45°), and then, as I go on, will decrease to a square again. That seems a promising start. I'll twizzle **C** until I'm looking at it along a line through two of its opposite edge-centres (when what I'll see will be that rectangle). Then I'll think again.

4   To fix their ideas, they labelled as E, N, W, S the vertices of one face of **C**, taken in order round the face; and labelled as E', N', W', S' the vertices of **C** lying diametrically opposite E, N, W, S, respectively.

They placed **C** (in relation to rectangular Cartesian coordinates) with its centre at the origin $(0, 0, 0)$, with E at $((\sqrt{2})n, 0, n)$, and with N at $(0, (\sqrt{2})n, n)$.

> (See Diagram 12.1. *They* are looking at **C** from the south, in the (X, Y)-plane; *we* are looking from the south-south-east, and we are above the (X, Y)-plane.)

Each of them now decided to rotate **C** through some angle about the

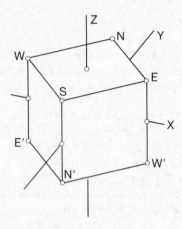

Diagram 12.1

X-axis; but at this stage they parted company in deciding what that angle should be.

5   *Tom*

Tom argued:

> When I now rotate **C** about the X-axis, what I'll see will be a hexagon. I want to make the largest possible hole in **C**, so it seems reasonable to use the largest possible hexagon. That will happen when I've rotated **C** to the stage in which the line SS′ is the line along which I'm looking. And that hexagon will be regular. This seems very plausible: the largest possible hexagon, regular hexagon, and an axis for my hole that's an axis of symmetry of **C**. It must be right.
>
> What I'll see will be a regular hexagon with sides $(2\sqrt{(\frac{2}{3})})n$ (Diagram 12.2(i)). Into it I can fit a square with sides $2\sqrt{2}(\sqrt{3}-1)n$. (I can do it in two different ways—see Diagrams 12.2(ii, iii)—but, since the hexagon is regular, they come to the same thing.) So I can pass **D** through **C** for any value of $\mu^2 < 4(2-\sqrt{3})$.
>
> In particular, when **C** has face area 36 square inches, I can do it for any **D** with face-area less than $144(2-\sqrt{3})$ square inches. Now $38 < 144(2-\sqrt{3}) < 39$. So I can do it with that 38-square-inch cube of Anne's, but not with either of the larger ones.

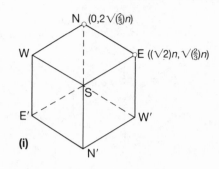

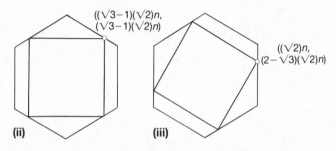

Diagram 12.2

6   *Dick*

Dick argued:

> I've just rotated **C** through 45° about the Z-axis, which seemed to be a good idea; so it now seems reasonable to give **C** another rotation through 45°—but this time about the X-axis.
>
> What I'll then see will be a hexagon (Diagram 12.3(i)).
>
> What's the largest square that I can fit into that hexagon? There seem to be two possibilities (see Diagrams 12.3(ii, iii)). One has area $4n^2$ (Diagram 12.3(ii)), but the other has area $2(5-2\sqrt{2})n^2$ (Diagram 12.3(iii)). So I can pass **D** through **C** for any value of $\mu^2 < \frac{1}{2}(5-2\sqrt{2})$.
>
> In particular, when **C** has face-area 36 square inches, I can do it for any **D** with face area less than $18(5-2\sqrt{2})$ square inches. Now $39 < 18(5-2\sqrt{2}) < 40$. So I can do it with that 39-square-inch cube of Anne's, but not with the larger one.

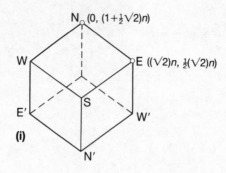

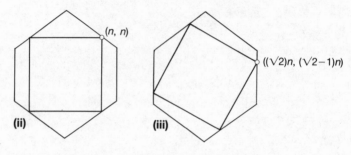

Diagram 12.3

7   *Harry*

Harry argued:

> When I now rotate **C** through an angle θ about the X-axis, that will move

E	from $((\sqrt{2})n, 0, n)$	to $((\sqrt{2})n, -n\sin\theta, n\cos\theta)$,
N	from $(0, (\sqrt{2})n, n)$	to $(0, (\sqrt{2})n\cos\theta - n\sin\theta,$
		$(\sqrt{2})n\sin\theta + n\cos\theta)$,
W	from $(-(\sqrt{2})n, 0, n)$	to $(-(\sqrt{2})n, -n\sin\theta, n\cos\theta)$,
S	from $(0, -(\sqrt{2})n, n)$	to $(0, -(\sqrt{2})n\cos\theta - n\sin\theta,$
		$-(\sqrt{2})n\sin\theta + n\cos\theta)$;

> and E′, N′, W′, S′ to new positions diametrically opposite the new positions of E, N, W, S.
>
> What I'll then see will be a hexagon (Diagram 12.4(i)).

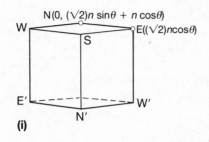

$N(0, (\sqrt{2})n\sin\theta + n\cos\theta)$

W
$E((\sqrt{2})n\cos\theta)$

S

E'
W'

N'

(i)

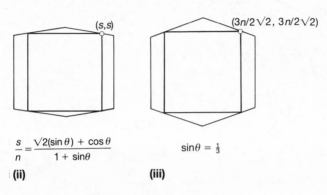

(s,s)

$(3n/2\sqrt{2}, 3n/2\sqrt{2})$

$$\frac{s}{n} = \frac{\sqrt{2}(\sin\theta) + \cos\theta}{1 + \sin\theta}$$

(ii)

$\sin\theta = \tfrac{1}{3}$

(iii)

Diagram 12.4

What is the largest square that I can fit into that hexagon? For relatively small values of θ, there's only one possibility (see Diagram 12.4(ii)): the square will have side-length $2s$, where

$$\frac{s}{n} \equiv \frac{s(\theta)}{n} = \frac{\sqrt{2}\sin\theta + \cos\theta}{1 + \sin\theta}.$$

As θ varies, $s(\theta)/n$ attains its maximum value when $\sin\theta = \tfrac{1}{3}$, so I'll take that value of θ. Then $s/n = 3/2\sqrt{2}$ (see Diagram 12.4(iii)); so I can pass a **D** through **C** for any value of $\mu^2 < \tfrac{9}{8}$.

In particular, when **C** has face-area 36 square inches, I can do it for any **D** with face area less than $40\tfrac{1}{2}$ square inches. So I can do it with any of Anne's cubes.

8 *Addendum*

Confining ourselves (as Tom, Dick, and Harry have done) to situations in which **C** is first rotated through $\tfrac{\pi}{4}$ about the Z-axis (as in

Section 3), and then rotated through θ about the X-axis, let $4\mu^2(\theta)n^2$ be the face-area of the largest **D** that will pass through a **C** having face area $4n^2$. $\mu^2(\theta)$ is sketched in Diagram 12.5.

$\mu^2(\theta)$ increases from 1 (at $\theta=0$) to a maximum $\frac{9}{8}$ (at $\sin\theta=\frac{1}{3}$ (Harry's solution)).

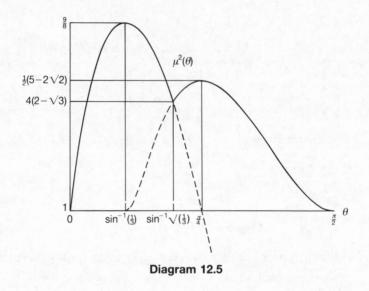

Diagram 12.5

For $0\leq\sin\theta<\frac{1}{3}$ there is only one square that we need to consider: the 'normal' one, with two of its sides parallel to the Z-axis. When $\sin\theta\geq\frac{1}{3}$, there are two squares that we need to consider; but the second one—the 'skew' one—has a smaller area than has the 'normal' one (for $\frac{1}{3}\leq\sin\theta\leq\sqrt{(\frac{1}{3})}$).

$\mu^2(\theta)$ then decreases to $4(2-\sqrt{3})$ (at $\sin\theta=\sqrt{(\frac{1}{3})}$ (Tom's solution—Tom in fact hit on a local minimum)).

At $\sin\theta=\sqrt{(\frac{1}{3})}$ the normal square (decreasing in area as θ increases) and the skew square (still increasing) have equal area: from now on the skew square takes over in providing the value of $\mu^2(\theta)$.

$\mu^2(\theta)$ then increases to a (local) maximum $\frac{1}{2}(5-2\sqrt{2})$ (at $\theta=\frac{\pi}{4}$ (Dick's solution)), and then decreases to 1 (at $\theta=\frac{\pi}{2}$).

Anne's intervention

Anne intervened: 'Harry's shown that one can make a hole in **C** through which one can pass a cube **D** if $\mu^2 < \frac{9}{8}$. But how about the converse—that one can't possibly do it[3] if $\mu^2 > \frac{9}{8}$? I know that I didn't ask anyone to prove the converse—but it would be a nice proof to have.'

'How about this for an argument?' said Belinda. 'First of all, think of a hollowed-out **C**—a cubical box with faces of infinitesimal thickness: let's call it **B**. Then think of an infinitesimally thin slice of **D**—one of its faces, in effect: let's call that lamina **L**. Can we put **L** in **B**? If we can, well and good: we can make a hole in **C**, with **L** as its cross-section, through which **D** can pass. But if we *can't* fit **L** into **B** in any way, then we can't make such a hole. It should be fairly easy to find out what's the largest square lamina that we can fit into a cubical box.'

'Good for you, Belinda,' said Tom, 'and I'm sure that it will have $\mu^2 = \frac{9}{8}$.'

'It has,' said Harry (who'd been working furiously on the back of a [large] envelope while Belinda was expounding): 'so that's the proof of your converse, Anne.'

'*Just* a moment,' said Anne. 'I agree that if **D** *can't* pass through **C**, then the lamina **L**—the face of **D**—can't fit inside **B**. But it doesn't follow that, if **D** *can* pass through **C**, then **L** can fit inside **B**.'

'I didn't say that it "followed",' said Belinda. 'I just took it as obvious. After all, if we can pass **D** through **C**, then the leading face (for example) of **D** has to lie inside **C** at some time.'

'Each point of the leading face of **D** has to lie within **C** at some time, I agree,' said Dick; 'but don't you have to show that all those points lie within **C** at the *same* time?'

'Precisely,' said Anne.

'Bother,' said Belinda. 'But I think that you're being too pedantic. After all, it *is* obvious, isn't it?'

'No,' said Anne. 'You've put it persuasively—but let's go back to those hexagons and squares that Tom, Dick, and Harry were talking about. But this time take the general case: twizzle **C** through ϕ (rather than their specific $\frac{\pi}{4}$) about the Z-axis, and then through θ about the X-axis. We'll see a hexagon with a square inscribed in it. We go on to

---

[3] I deliberately ignore the case $\mu^2 = \frac{9}{8}$. (Whether to include it in the 'can' or in the 'can't' category depends entirely on one's views on 'when is a hole not a hole?')

find—for varying θ and ϕ—the situation in which the size of the square is a maximum. The four points at which the square touches the hexagon are, in the two-dimensional projection we're looking at, the corners of a square (by definition). But they're also the projections of the four points at which edges of **D** touch edges of **C**. Is it obvious that *those* four points are even coplanar—never mind being the corners of a square?'

'No,' said Belinda, 'it isn't.'

'But that's the point that Dick's making, isn't it?'

'Double bother,' said Belinda.

Models

1   Diagram 12.6 shows what is left of **C** (of side $2n$) after a hole has been made in it that will accommodate a **D** (of side $2\mu n$): let us call this residual solid R_μ.

> If $\mu^2 < \frac{9}{8}$ one can make an infinity of such holes. I have chosen to use 'Harry's axis'—a line, **A**, through the centre of **C**, in the plane NSN'S', making an angle $\sin^{-1}(\frac{1}{3})$ with NS. The hole has **A** as axis, and has as cross-section a square (of side-length $2\mu n$) two of whose sides are parallel to WE.
>
> In Diagram 12.6, unbroken lines are visible edges of R_μ; broken lines are hidden edges of R_μ; and dotted lines have no physical reality—they are just useful construction lines.

2   One obvious way to make a model of R_μ is to start with a solid cube **C** (or the necessary part of one) and to bore through it the appropriate hole. That, however, requires relatively sophisticated equipment.

3   A simpler method is to make four separate (convex) solid blocks—illustrated in Diagram 12.7—and then to glue them together. The drawback here is that each of the outer faces of the model will have one or two 'junction marks', where two blocks join.

4   The simplest method—the one using the least equipment—is to make a (hollow) model, using cardboard and glue. One way of cutting and folding the cardboard to form the model is illustrated in Diagram 12.8. The difficulty here is that the thickness of the cardboard generates significant complications—at least, it does so for 'interesting' values of μ (> 1).

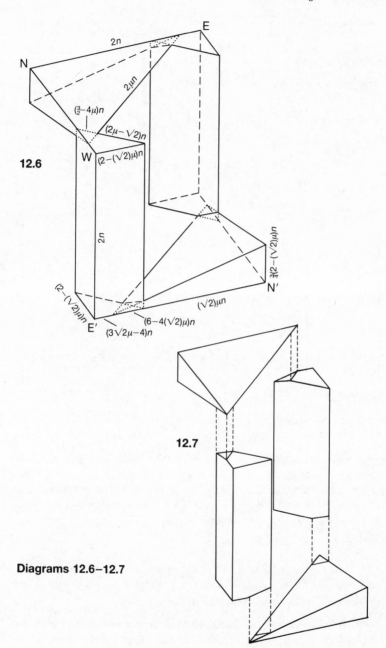

12.6

$2n$

N

E

$2\mu n$

$(\frac{3}{2}-4\mu)n$

$(2\mu-\sqrt{2})n$

W $(2-(\sqrt{2})\mu)n$

$2n$

$(2-(\sqrt{2})\mu)n$

$\frac{3}{4}(2-(\sqrt{2})\mu)n$

N'

$(\sqrt{2})\mu n$

$(6-4(\sqrt{2})\mu)n$

E' $(3\sqrt{2}\mu-4)n$

12.7

Diagrams 12.6–12.7

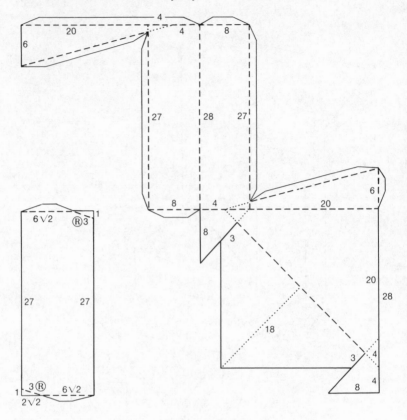

Diagram 12.8

Cut along the unbroken lines. Fold along the broken lines (the two lines marked ® should be folded in the reverse sense). The dotted lines are construction lines only (they should not be cut or folded, and should be erased before assembly). You need one each of the two pieces sketched here, and one each of their mirror-images.

5   *All* methods share an intrinsic problem (physical weakness in four regions of the model), which is perhaps best illustrated in the context of the method described in para. 3.

6   R_μ will—in theory—hold together for any $\mu^2 < \frac{9}{8}$; but the area of each of the 'glueing areas' connecting the four blocks described in

para. 3 is $\mathbf{G}_\mu$:

$$\mathbf{G}_\mu = \frac{3}{2\sqrt{2}} \, (3 - (2\sqrt{2})\mu)^2 n^2.$$

When $n = 3$ in. (as is the case with the **C** that Anne presented) and $\mu = \frac{10}{9}$ (as is the case with the **D** that Harry could pass through **C**),

$$\mathbf{G}_\mu = \frac{3}{2\sqrt{2}} \, (9 - 4\sqrt{5})^2 \text{ sq. in.} \cong 0.0033 \text{ sq. in.}$$

so that to make his hole Harry will require considerable precision (and will also need to handle his eventual $\mathbf{R}_{10/9}$ delicately).

Still with $n = 3$ in., less ambitious, but less impractical, are models with

$$\mu^2 = \tfrac{50}{49} \qquad (\mathbf{G}_\mu = \tfrac{27}{98\sqrt{2}} \text{ sq. in.} \cong 0.195 \text{ sq. in.})$$

or

$$\mu^2 = \tfrac{289}{288} \qquad (\mathbf{G}_\mu = \tfrac{3}{8\sqrt{2}} \text{ sq. in.} \cong 0.265 \text{ sq. in.}).$$

> I suggest these particular values of μ because with them nearly all the lengths that one has to measure off in making the model of $\mathbf{R}_\mu$ are rational multiples of each other.

(Diagrams 12.6 and 12.7 (and the frontispiece of this book) are drawn with $\mu^2 = \frac{289}{288}$. Diagram 12.8 is drawn with $\mu^2 = \frac{50}{49}$.)

Composer's problem

In 1987 I came across a note that I had written in 1968—a few lines of typescript and a couple of diagrams—that blandly asserted (translated into the notation that I have used here) that 'cube through cube' was possible if and only if $\mu^2 < \frac{9}{8}$.

'Aha,' I said to myself, 'a few days' work to re-create the proof (which I seem not to have kept), and to put the problem into context; and another day or two to make a decent model.'

It took me a lot longer than that.

There were two difficulties to be overcome, of different kinds. Since I haven't yet overcome either of them, I prefer to transfer them as rapidly as possible to the Extension section.

Extension

I have two extension questions.

1   We know that 'cube through cube' is possible if $\mu^2 < \frac{9}{8}$. Belinda attempted a proof that it is impossible if $\mu^2 > \frac{9}{8}$, but failed; and the direct analytic approach seems rather fearsome. So:

> Can you produce a fairly short, rigorous, proof that 'cube through cube' is impossible if $\mu^2 > \frac{9}{8}$?

2   I would have liked to make a model **R** that could be kept in a box **D** (and, of course, such that **D** could pass through **R**). My engineering skills aren't up to it.

Are yours?

13

POTENTIAL PAY

This is not a 'Problem'—in the sense that nearly all the other chapters in this book involve 'problems'. It's a comment on a classic problem.

Interested, despite himself, in the salaries advertised for semi-numerate jobs in the City, Tom ventured there for a couple of interviews. Returning to the relative sanity and security of the Ayling Arms, he reported:

'As far as I could make out, the jobs were identical. But Buy-Em-And-Strip-Them offered £20,000 a year as starting salary, with a rise of £4,000 every year; and What's-His-Is-Mine offered £20,000 a year as starting salary, with a rise of £1,000 every half-year. Now, which would you take?'

'It's obvious,' said Dick: 'BEAST's. I admit that you'd do a bit better with WHIM during the second half of the first year; but after that with BEAST you'd race ahead.'

'Hang on,' said Harry; 'this is a classic—don't you remember those two clerks in Rouse Ball?[1] In the first year you'd get £20,000 with BEAST, but £10,000 + £11,000 with WHIM; in the second year you'd get £24,000 with BEAST, but £12,000 + £13,000 with WHIM; in the third year you'd get £28,000 with BEAST, but £14,000 + £15,000 with WHIM. And so on. Each year you'd get £1,000 more with WHIM than with BEAST. You ought to have realized that: after all, the Two Clerks Paradox has been knocking around for ages.'

'I know it's a classic paradox,' said Anne. 'It comes up once every twenty or thirty years or so—and it's high time that it got knocked on the head for good and all. You're right, Harry, when you say that as far as BEAST is concerned Tom would get £20,000 in the first year, £24,000 in the second year, £28,000 in the third year, and so on. But

---

[1] W. W. Rouse Ball, *Mathematical Recreations and Essays*, 11th edn. (Macmillan, 1939), p. 47 (cited *in extenso* in 'Discussion' section).

you've got it all wrong about what he could expect from WHIM. Let's take it step by step. Tom starts at a rate of pay of £20,000 a year. So he gets £10,000 for his first half-year's work. Then he gets 'a rise of £1,000'. So what's his rate of pay during his second half-year's work?'

'£21,000.'

'Let's be a bit more precise: a rate of pay is a rate of pay, not a sum of money.'

'Oh, all right,' said Harry: '£21,000 a year.'

'So what does Tom get for his second half-year's work?'

'£10,500, I suppose,' said Harry.

'Right. Not the £11,000 that you said before. Then he gets another "rise of £1,000". So his rate of pay during his third half-year's work is £22,000 a year. So for that half-year's work he gets £11,000. And so on. Let's tabulate it.' (See Table 13.1.)

Table 13.1

| | With BEAST | | With WHIM | |
	Rate of pay	Earnings	Rate of pay	Earnings
1st half-year	£20,000 p.a.	£10,000	£20,000 p.a.	£10,000
2nd half-year	£20,000 p.a.	£10,000	£21,000 p.a.	£10,500
3rd half-year	£24,000 p.a.	£12,000	£22,000 p.a.	£11,000
4th half-year	£24,000 p.a.	£12,000	£23,000 p.a.	£11,500
5th half-year	£28,000 p.a.	£14,000	£24,000 p.a.	£12,000
6th half-year	£28,000 p.a.	£14,000	£25,000 p.a.	£12,500

'So where', said Harry, 'did I go wrong? And, if I'm wrong, why has the Two Clerks Paradox been around for so long?'

'That', said Anne firmly, 'is something we can discuss later. But first of all I want to hear from Tom—what's he going to do?'

'Didn't I tell you?' said Tom: 'Turn them both down, of course. It's much more fun here—and I can't see Anne being the wife of a Yuppie . . .'

'Is that a proposal?' said Anne.

'That', said Tom firmly, 'is something we can discuss later.'

Discussion

The Two Clerks Paradox, as described in Rouse Ball's classic book,[2] is:

> Two clerks, A and B, are engaged, A at a salary commencing at the rate of (say) £100 a year with a rise of £20 every year, B at a salary commencing at the same rate of £100 a year with a rise of £5 every half-year, in each case payments being made half-yearly: which has the larger income? The answer is B; for in the first year A received £100, but B receives £50 and £55 as his two half-yearly payments, and thus receives in all £105. In the second year A receives £120, but B receives £60 and £65 as his two half-yearly payments and thus receives in all £125. In fact, B will always receive £5 a year more than A.

The reason that Harry—and Rouse Ball—went wrong, and the reason that the Two Clerks Paradox has been around so long, is, simply, *Misuse of Units*.

If Tom had reported 'BEAST offered £20,000 a year as starting salary, with a rise of £4,000 a year every year; and WHIM offered £20,000 a year as starting salary, with a rise of £1,000 a year every half-year', then Harry would (almost certainly) not have gone wrong. But Tom's report was in colloquial form, referring to a 'rise of £1,000' rather than the correct 'rise of £1,000 a year'; and that led Harry to assume that Tom would get—with WHIM—£1,000 more each half-year than he had done in the previous half-year.

It may be protested that that is what WHIM in fact offered. I cannot disprove it. (One cannot disprove a meaningless assertion—and the concept of 'a rise of £1,000' is as meaningless as the concept of 'a speed of 10 miles'.) But I ask you: just imagine the response that Tom would have got (supposing that he had joined WHIM) at the end of the following dialogue:

WHIM: 'Tom, you've now been with us for six months at £20,000 a year; so we're giving you a rise of £1,000.'

Tom: 'Thank you. So that means that for the next six months I'll be paid at the rate of £22,000 a year?'

WHIM: ! . . . !

Composer's problem

My Physics master reserved his greatest ire for those who confused weight with mass. But *any* work submitted to him that used

[2] Rouse Ball's *Mathematical Recreations and Essays* was first published in 1892. But the context in which he places the statement of the Two Clerks Paradox clearly suggests that he is quoting from a much earlier source (which unfortunately he doesn't cite—or, at least, doesn't cite in my (1939, 11th) edition).

anomalous units (I do not mean, for example, 'feet' instead of 'metres': I mean 'feet' instead of 'feet per second') came back with **UNITS PLEASE** stamped all over it; scored zero marks; and had to be done again from scratch (out of school hours).

Rumour had it—I cannot vouch for it—that he used the same **UNITS PLEASE** stamp daily on the financial pages of his newspaper before they were returned to the Editor accompanied by scathing comments on the innumeracy of the Financial Editor. (Unfortunately, there can have been little impact: 45 years later it still seems to be the rule rather than the exception for financial journalists to use '%' where '% per annum' is intended.[3])

The misuse of units is always irritating (the more so because it is so unnecessary a sloppiness). Sometimes it leads to considerable confusion. Occasionally it leads to a completely incorrect result.

I have a number of relatively complex examples of completely incorrect results.[4] But I wanted a simple example. Unfortunately, the only really simple one that I have been able to find is the Two Clerks Paradox.

---

[3] Interestingly, their non-financial colleagues are less culpable: motoring correspondents seldom if ever refer to a car as having a top speed of 130 miles.

[4] One that I would have liked to have used arose in a financial–legal case in which I had been engaged as an expert witness. The case depended—*inter alia*—on the difference between '26% a year' and '26% each year'. (Compare '26 miles an hour' and '26 miles each hour'.) The judge could see no difference between them. (Luckily, Lord Denning MR was sitting when the case came to the Court of Appeal—and he is a mathematician.) But I have not been able to refine the incident into a simple straightforward example.

14

BENEDICT'S BIRTHDAY[1]

Problem

Archibald and Angela are brother and sister, and their ages add up to a perfect square; Benedict and Beatrice are brother and sister, and their ages add up to a perfect square; Cuthbert and Constance are brother and sister, and . . . —but I won't go through the whole list of them: there are, in all, thirteen of these brother–sister pairs, each pair with ages adding up to a perfect square.

No two of the twenty-six are the same age. Archibald, the oldest, is 26; Benedict, the youngest, is 1. The ages of Cuthbert and Constance differ by 5.

All of them sit down at a circular table, brother opposite sister, to celebrate Benedict's birthday.

Each has a neighbour who is 5 years older or younger.

Both of his neighbours are younger than Cuthbert.

What are the ages in order round the table (starting with Angela and then the younger of her two neighbours)?

[1] First published on 21 August 1988 as Brainteaser 1355 in *The Sunday Times Magazine*.

Solution

1 *Who is opposite whom?*

We want to arrange the integers $1, 2, \ldots, 26$ as thirteen pairs, with the sum of each pair a perfect square. (There is more than one way in which this can be done, so at an early stage we shall use the additional restraint that the members of one of the pairs—Cuthbert and Constance—differ by 5.)

What can pair with 18? Only 7.	So: $18 + 7$
What ($\neq 7$) can pair with 9? Only 16.	$9 + 16$
What ($\neq 16$) can pair with 20? Only 5.	$20 + 5$
$\lvert\text{Cuth-Con}\rvert = 5$. Cuth, Con $\neq 7$.	$15 + 10$ (C + C)
What ($\neq 10$) can pair with 26? Only 23.	$26 + 23$ (Arch + Ang)
What ($\neq 15$) can pair with 21? Only 4.	$21 + 4$
What ($\neq 7, 23$) can pair with 2? Only 14.	$2 + 14$
What ($\neq 14$) can pair with 22? Only 3.	$22 + 3$
What ($\neq 5, 14$) can pair with 11? Only 25.	$11 + 25$
What ($\neq 3, 23$) can pair with 13? Only 12.	$13 + 12$
What ($\neq 3, 10$) can pair with 6? Only 19.	$6 + 19$
What ($\neq 19$) can pair with 17? Only 8.	$17 + 8$
What ($\neq 12, 25$) can pair with 24? Only 1.	$24 + 1$ (Bea + Ben)

So each integer determines its pair uniquely.

2 *Who neighbours whom?*

Each integer must have a neighbour that differs from it by 5. It is convenient to start with 26 (Archibald).

$$\text{Neighbours} \begin{Bmatrix} 26 \\ 23 \end{Bmatrix} \text{so} \begin{Bmatrix} 21 & 26 \\ & 23 \end{Bmatrix} \text{so} \begin{Bmatrix} 21 & 26 & \\ 4 & 23 & 18 \end{Bmatrix} \text{so} \begin{Bmatrix} & 21 & 26 & 7 \\ 9 & 4 & 23 & 18 \end{Bmatrix} \text{so} \ldots$$
$$\text{Opposite}$$

$$\text{Neighbours} \quad \text{so} \begin{Bmatrix} 10 & 5 & 11 & 16 & 21 & 26 & 7 & 2 & 6 & 1 \\ 15 & 20 & 25 & 9 & 4 & 23 & 18 & 14 & 19 & 24 \end{Bmatrix} \text{and} \begin{Bmatrix} 22 & 17 & 12 \\ 3 & 8 & 13 \end{Bmatrix}$$
$$\text{Opposite}$$

Cuthbert is 10 or 15. But both of his neighbours are younger than he. So he is not 15 (since we know that the 15-year-old has a neighbour of 20). So Cuthbert is 10 (with the 5-year-old as one of his neighbours). His other neighbour must be 3, 12, 13, or 22; and, so, must be 3. Hence

Neighbours	13	8	3	10	5	11	16	21	26	7	2	6	1
Opposite	12	17	22	15	20	25	9	4	23	18	14	19	24

3   *The order round the table*
It follows that the order round the table (starting with 23 (Angela)
and then 4 (the younger of her neighbours)) is: 23, 4, 9, 25, 20, 15, 22,
17, 12, 1, 6, 2, 7, 26, 21, 16, 11, 5, 10, 3, 8, 13, 24, 19, 14, 18.